Grade 2

MATH INTERVENTION
Geometry, Measurement & Data

hand2mind®

Hands-On Standards®
TEACHING MATH WITH MANIPULATIVES

Online Resources

To access the online resources,
visit **https://online.hand2mind.com/**. Enter your
email address and password or click
"New user? **Register here**" to set up an account.

Once you log in, enter your access code:

8QDD-DMWK-HVO3-J1VO

Hands-On Standards® Math Intervention: Geometry, Measurement & Data, Grade 2
93492
978-1-63406-793-5

hand2mind.

500 Greenview Court • Vernon Hills, Illinois 60061-1862 • 800.445.5985 • **hand2mind.com**

© 2021 hand2mind, Inc., Vernon Hills, IL, USA
All rights reserved.

No part of this publication may be reproduced, stored in a retrieval system, or transmitted, in any form or by any means, electronic, mechanical, photocopying, recording, or otherwise, without the prior written permission of the publisher. Permission is granted for limited reproduction of those pages from this book with a copyright line in the outer margin for classroom use and not for resale.

Printed in the United States of America.
H1208210311

Contents

Introduction..1
A Walk Through a Lesson......................2
A Walk Through Student Pages...........4

Geometry 6
Lesson 1 Identify Shapes........................8
Lesson 2 Recognize and Draw Shapes............10
Lesson 3 Partition Rectangles............12
Lesson 4 Solve Problems by Partitioning Rectangles..14
Lesson 5 Partition Rectangles into Fair Shares......16
Lesson 6 Partition Circles...................18

Measurement 20
Lesson 1 Estimate Lengths..................22
Lesson 2 Different Size Units..............24
Lesson 3 Select and Use Measurement Tools......26
Lesson 4 Measure and Compare Lengths..........28
Lesson 5 Whole Numbers as Lengths on a Number Line......30
Lesson 6 Tell Time to 5 Minutes..........32
Lesson 7 Tell Time to the Minute.........34
Lesson 8 Solve Coin Problems.............36

Data 38
Lesson 1 Line Plots Using Inches........40
Lesson 2 Line Plots Using Centimeters...........42
Lesson 3 Solve Problems Using a Line Plot......44
Lesson 4 Picture Graphs........................46
Lesson 5 Bar Graphs..............................48
Lesson 6 Solve Problems Using Graphs...........50

Lesson Student Pages

Geometry 52
SP Lesson 1 Identify Shapes.................52
SP Lesson 2 Recognize and Draw Shapes..........54
SP Lesson 3 Partition Rectangles..........55
SP Lesson 4 Solve Problems by Partitioning Rectangles......56
SP Lesson 5 Partition Rectangles into Fair Shares....57
SP Lesson 6 Partition Circles...............58

Measurement 59
SP Lesson 1 Estimate Lengths...............59
SP Lesson 2 Different Size Units..........61
SP Lesson 3 Select and Use Measurement Tools......63
SP Lesson 4 Measure and Compare Lengths.......65
SP Lesson 5 Whole Numbers as Lengths on a Number Line......67
SP Lesson 6 Tell Time to 5 Minutes......70
SP Lesson 7 Tell Time to the Minute.....71
SP Lesson 8 Solve Coin Problems..........72

Data 74
SP Lesson 1 Line Plots Using Inches......74
SP Lesson 2 Line Plots Using Centimeters..........76
SP Lesson 3 Solve Problems Using a Line Plot......78
SP Lesson 4 Picture Graphs....................80
SP Lesson 5 Bar Graphs..........................82
SP Lesson 6 Solve Problems Using Graphs........84

Multi-Lesson Blackline Masters

BLM 1 1-Inch Grid Paper 86
BLM 2 Telling Time Recording Sheet 87

Assessments

Geometry 88
Measurement 91
Data 93
Assessment Student Progress Report 97

Answer Key 99
Glossary of Manipulatives 105
Index 107

Introduction

How do we help students find meaning in mathematics? That is, how do we give students more than a rote script for reciting facts and churning out computations? How do we help students develop conceptual understanding and fluency?

Hands-On Standards® Math Intervention: Geometry, Measurement & Data, Grade 2 is an easy-to-use reference manual for teachers who want to help students discover meaning in mathematics. Each of the manual's 20 lessons demonstrates a hands-on exploration using manipulatives. The goal is to help students get a physical sense of a problem—to help students get their hands on the concepts they need to know and to help them "see" the meaning in mathematics.

Each lesson in *Hands-On Standards® Math Intervention: Geometry, Measurement & Data, Grade 2* targets a clearly stated objective. The main part of a lesson offers a word problem that students can relate to and has the student work on the problem using a hands-on approach. Full-color photographs demonstrate the suggested steps. In addition to the main activity, each lesson includes suggested points of discussion, ideas for more exploration, a formative assessment item, English language support, and practice pages to help students solidify their understanding. The instructional model is a progression from concrete to abstract.

This book is divided into three sections—Geometry, Measurement, and Data. At the end of each section, a Unit Assessment and Progress Monitoring tool are provided in order to evaluate and track students' understanding of the content.

Each lesson in this book uses one or more of the following manipulatives:

- Coin Tiles
- Color Tiles
- Cuisenaire® Rods
- Rainbow Fraction® Circles
- Geoboard

- Inchworms™
- Inchworms™ and Centibugs™ Ruler
- Mini Relational GeoSolids®
- NumberLine Clock™

Read on to find out how *Hands-On Standards® Math Intervention: Geometry, Measurement & Data, Grade 2* can help the students in your class find meaning in math and build a foundation for future math success!

A Walk Through a Lesson

Each lesson in *Hands-On Standards Math Intervention: Geometry, Measurement, and Data, Grade 2* includes many features, including background information, objectives, pacing and grouping suggestions, discussion questions, and ideas for further activities, all in addition to the step-by-step hands-on activity instruction. Take a walk through a lesson to see an explanation of each feature.

Lesson Introduction
A brief introduction explores the background of the concepts and skills covered in each lesson. It shows how they fit into the larger context of students' mathematical development.

Try It!
The **Try It!** section provides a hands-on activity for students to explore the mathematical concept using a real-world problem.

Objective
The **Objective** summarizes the skill or concept students will learn through the hands-on lesson.

Materials
The **Materials** box lists the type and quantity of materials that students will use to complete the activity, including manipulatives, lesson-specific student pages, and multi-lesson blackline masters.

EL Support
EL Support provides suggestions for ways that teachers can provide English language lesson support. Additionally, consider how these supports might be used for all learners when needed.

Lesson 1 Line Plots Using Inches

Students at this age have learned how measure lengths. Here, students use that skill to measure the length of objects and plot these on a line plot. This will lay the foundation for students to create and plot other types of graphs.

Perform the **Try It!** activity on the next page.

Objective
Measure and graph the length of four objects in inches using a line plot.

Materials
- Inchworms™ (10 per pair)
- Bookmark Line Plot Recording Sheet (Lesson 1, page 74, 1 per pair)

EL Support
- Review vocabulary: line, plot.
- Discuss the everyday meanings of the words *line* and *plot*. Have the students give examples of lines such as when the class lines up for lunch or a line on a road. Explain to students that a plot can be a piece of land or what happens in a story.
- Write the following sentence frame to be used during the **Try It!** I can find the length by _____.

Talk About It
Discuss the **Try It!** activity.

- **Ask:** *What does each X represent?* Students should understand that each X represents one object with that length.
- **Ask:** *What can you tell from comparing the height of the different columns?* Students should explain that the height of the plot above each measurement is the number of things with that length. By looking at the heights of the different columns, they can quickly compare the number of things with that height.
- **Ask:** *What are some other things you might place on a line plot?* Students may mention other measurements such as the lengths of nails in a tool kit or the number of glasses a restaurant has in each size.

Solve It
With the students, reread the problem. **Ask:** *What was the length of each bookmark? What did the line plot show?*

More Ideas
For other ways to teach measuring and creating a line plot—

- Have students use Color Tiles to measure classroom objects such as a crayon, a pencil, and paper. Then have students use that data to create a line plot.
- To create a line plot, give the students a sticky note with an X on it. Create a number line from 0 to 10 on the board. Have the students place their sticky note above the number that represents the number of siblings they have. Discuss the findings.
- For more practice, use Lesson 1 student page 75.

40

Talk About It
The **Talk About It** section provides discussion topics and questions that can be used during or after the **Try It!** activity. This helps students connect the hands-on activity to a larger mathematical concept while building deeper understanding.

Solve It
Solve It gives students a chance to show what they've learned. Students are asked to return to and solve the original word problem. They might summarize the lesson concept through representation or writing, or extend the skill through a new variation on the problem.

More Ideas
More Ideas provides additional activities and suggestions for teaching about the lesson concept using a variety of manipulatives. These ideas might be suggestions for additional practice with the skill or an extension of the lesson.

Try It!

The **Try It!** activity opens with **Pacing** and **Grouping** guides. The **Pacing** guide indicates about how much time it will take for students to complete the activity, including the post-activity discussion. The **Grouping** guide recommends whether students should work independently, in pairs, or in small groups.

Next, the **Try It!** activity is introduced with a real-world word problem. Students will "solve" the problem by performing the hands-on activity. The word problem provides a context for the hands-on work and the lesson skill.

This section of the page also includes any instruction that students may benefit from before starting the activity, such as a review of foundational mathematical concepts or an introduction to new ones.

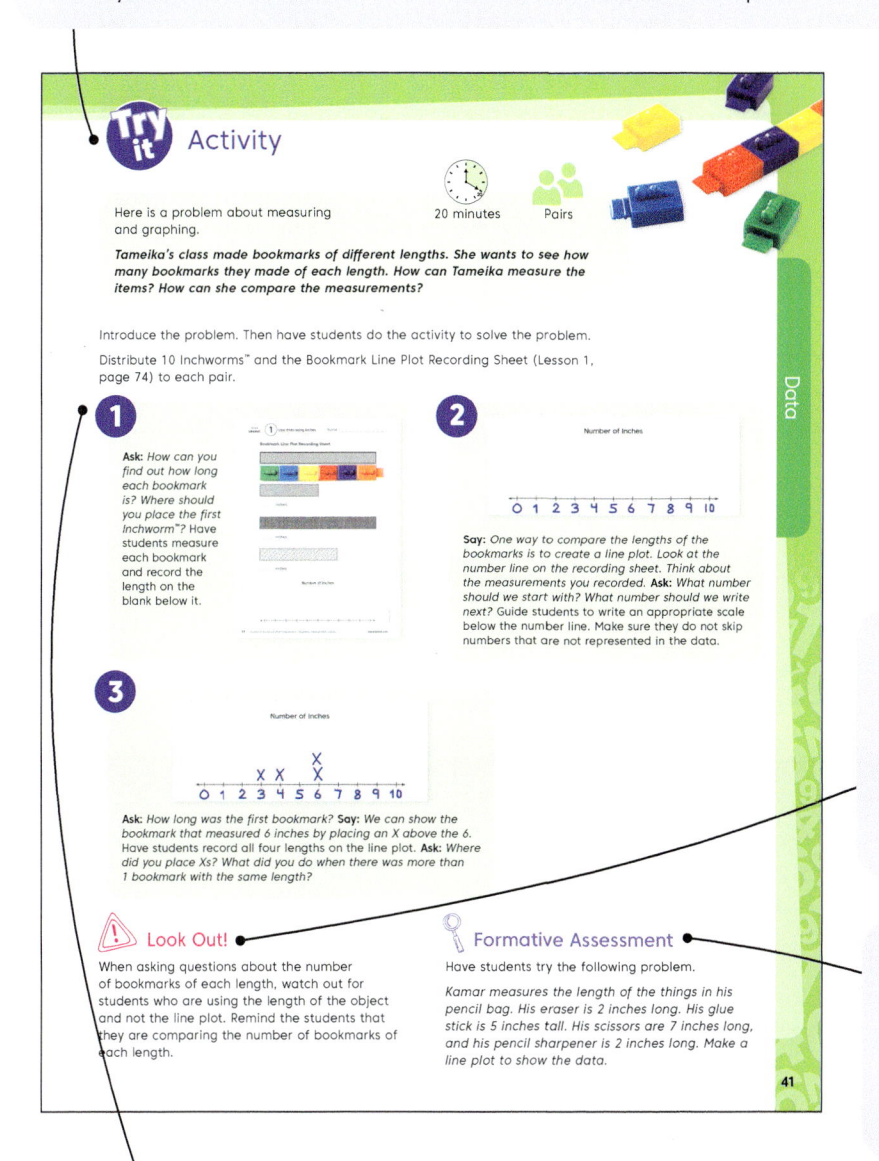

Look Out!

Look Out! describes common errors or misconceptions likely to be exhibited by students at this age dealing with each skill or concept and offers troubleshooting suggestions.

Formative Assessment

Formative Assessment provides a problem that allows for ongoing feedback on students' understanding of the concept.

Step-by-Step Activity Procedure

The hands-on activity itself is the core of each lesson. It is presented in three—or sometimes four—steps, each of which includes instruction in how students should use manipulatives and other materials to address the introductory word problem and master the lesson's skill or concept. An accompanying photograph illustrates each step.

A Walk Through a Lesson

A Walk Through Student Pages

The lessons are followed by a set of student pages that provide additional practice for each lesson. These pages take the student from the concrete to the abstract, completing the instructional cycle. Students begin by using manipulatives, move to creating visual representations, and then complete the cycle by working with abstract mathematical symbols.

Topic and Lesson Number
Lesson student pages are organized by topic and lesson number.

Lesson Title
The Lesson Title indicates which lesson the student page should be used with.

Exercise
Concrete and **Representational** exercises (pictorial representations of the featured manipulative) help students bridge conceptual learning to symbolic mathematics.

Standards-Based Math Practice
Abstract exercises provide standards-based math practice to allow students to deepen their understanding of the featured skill.

Extended Response
Extended Response exercises feature an open-ended constructed response question to help teachers gauge students' understanding.

Data Lesson **1** Line Plots Using Inches   Name _____

Use the Inchworms™ to measure each pencil.

1. _____ inches

2. _____ inches

3. _____ inches

4. _____ inches

Use the lengths of the pencils to complete the line plot.

5.   Number of Inches

0 1 2 3 4 5 6 7 8 9 10

Challenge! Maya unwraps a new pack of 4 pencils. What would a line plot of the lengths of these pencils look like?

Hands-On Standards Math Intervention: Geometry, Measurement & Data   hand2mind.com   75

Lesson-Specific Student Pages

Lesson-specific student pages help guide the student through a **Try It!** activity. They include recording sheets, work spaces to place manipulatives, and other key information needed for a specific lesson.

Multi-Lesson Blackline Masters

Multi-lesson blackline masters are general tools, such as place-value charts, that can be used across multiple lessons. These resources help students organize their work and answers when completing **Try It!** activities.

Assessments

Assessments may be used to check students' understanding of each topic.

Assessment Student Progress Report

Assessment student progress reports provide a way to keep track of student mastery of each concept in the topic.

A Walk Through Student Pages

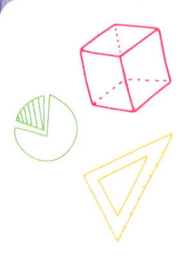

Geometry

In second grade, students describe and analyze shapes. By creating and analyzing two- and three-dimensional shapes, students develop a foundation for understanding geometry concepts such as congruence, similarity, and symmetry, which are necessary for learning in later grades. At this level, students recognize and draw shapes having specified attributes, such as a given number of sides or vertices. They analyze three-dimensional solids to identify attributes such as faces, edges, and vertices.

Students partition rectangles into rows and columns of same-size squares and count to find the total number of them. They solve word problems by identifying the total number of squares using a repeated addition number sentence.

Students also partition circles and rectangles into two, three, or four equal shares (or regions) and describe the shares by indicating how many equal shares are in halves, thirds, and fourths.

Geometry specifies that students should—

Reason with shapes and their attributes.

The following hands-on activities with manipulatives will help students grasp the geometry concepts presented in second grade. Mathematically proficient second-graders accurately use definitions and language to construct viable arguments about mathematics. During discussions about geometry problems, students should be given opportunities to constructively critique strategies and reasoning with their classmates. Teachers will want to ensure there is ample time for **students to communicate about shapes and their attributes.**

Geometry

Contents

Lesson 1 Identify Shapes 8
    Objective: Identify triangles, quadrilaterals, pentagons, and hexagons.
    Manipulative: Geoboard

Lesson 2 Recognize and Draw Shapes 10
    Objective: Recognize and draw shapes having specified attributes.
    Manipulative: Geoboard, Mini Relational GeoSolids®

Lesson 3 Partition Rectangles 12
    Objective: Partition rectangles into rows and columns.
    Manipulative: Color Tiles

Lesson 4 Solve Problems by Partitioning Rectangles 14
    Objective: Solve problems by partitioning a rectangle into rows and columns.
    Manipulative: Color Tiles

Lesson 5 Partition Rectangles into Fair Shares 16
    Objective: Partition a rectangle into rows and columns and describe shares using the words halves, thirds, half of, a third of, etc.
    Manipulative: Geoboard

Lesson 6 Partition Circles 18
    Objective: Partition circles into two, three, or four equal shares and describe the whole as two halves, three thirds, or four fourths.
    Manipulative: Rainbow Fraction® Circles

Lesson 1 Identify Shapes

Objective
Identify triangles, quadrilaterals, pentagons, and hexagons.

Materials
- Geoboard (1 per pair)
- Dot Paper (Lesson 1, page 52, 1 sheet per pair)

EL Support
- Review vocabulary: triangles, quadrilaterals, pentagons, hexagons. Identify the prefixes *tri-*, *qua-*, *pent-*, and *hex-* and discuss their meanings.
- Have students identify other words with the prefixes, such as *triplets* and *tricycles* for *tri-* and *quarter* for *qua-*. Show pictures and identify the parts, such as three wheels and four quarters in a dollar.
- Identify any cognates, such as triangle/el triángulo, quadrilateral/el cuadrilateral, pentagon/el pentágonon, hexagon/el hexágono.

Students at this stage have learned simple shapes such as triangles, circles, and rectangles. Here, students focus on identifying and naming shapes by counting the number of sides on a polygon. In later grades, students will build on this understanding when computing area and perimeter.

Perform the **Try It!** activity on the next page.

Talk About It

- Have students explore triangles. **Ask:** *What is a shape that can be made with three sides?* Display the representations of different triangles. **Ask:** *How are these shapes all the same? What are these shapes called?* Point out that while the triangles may look different, they are all classified as triangles because they have three sides.

- Display Geoboards that have various quadrilaterals such as squares, rectangles, rhombuses, and trapezoids. **Ask:** *How are all these shapes the same? What is a name that describes all of these shapes?* Point out that a quadrilateral describes all shapes with four sides.

Solve It

With students, reread the problem. **Ask:** *What is the name of the shape of Kira's tile? Why is it is called a hexagon?* Have students explain how they know the shape is a hexagon.

More Ideas

For other ways to teach about shapes—

- Use Pattern Blocks to reinforce identifying shapes. Have students sort the blocks into groups according to the number of sides. Then, ask students to identify the name of the shapes in each group. Point out that the trapezoid, parallelogram, and square are all quadrilaterals.

- Use the Geocircle side of the Geoboard to have the students create hexagons, triangles, and squares. Challenge the students to work with a partner to make a three-sided, four-sided, and six-sided shape. Then, ask students to name each shape.

- For more practice, use Lesson 1 student page 53.

Try it Activity

Here is a problem about shapes.

Kira is choosing tiles for her room. She wants them in the shape shown. What shape does Kira want? How can you find out?

15 minutes • Pairs

Introduce the problem. Then have students do the activity to solve the problem.
Distribute Geoboards and Dot Paper (Lesson 1, page 52) to students.

1
Ask: *How can Kira describe the shape of the tile?* Discuss the shape, guiding students to identify the number of sides. **Ask:** *How can you count the number of sides? What shape has this number of sides?*

2
Say: *Work with your partner. Make the shape on your Geoboard.* Then, have the students compare the shape they created with the shape that another group created. **Ask:** *Did all the hexagons look the same? Why might some look different? In what way are they all the same?*

3
Say: *Use dot paper to draw a picture of a hexagon that Kira could show to someone at a tile store.* Allow students to draw their shapes.

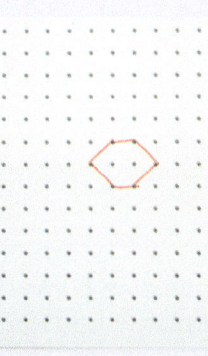

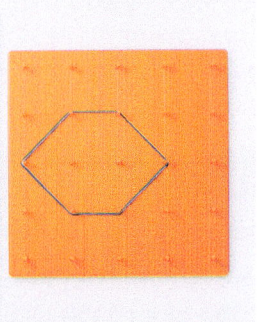

⚠ Look Out!

When sorting and categorizing shapes, watch out for students who do not recognize four-sided figures that are not squares or rectangles as quadrilaterals. You may wish to challenge the students to create as many different four-sided shapes on the geoboard as possible and guide the students to understand that each one is a quadrilateral.

🔍 Formative Assessment

Have students try the following problem.

Kira is looking at tiles in the store. She sees the tile below. How can she describe it? What name could she call it?

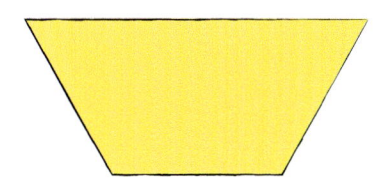

Geometry

9

Lesson 2 Recognize and Draw Shapes

Objective
Recognize and draw shapes having specified attributes.

Materials
- Geoboard (1 per pair)
- Mini Relational GeoSolids® (1 set per pair)

EL Support
- Review vocabulary: cube, prism, cone, pyramid, cylinder.
- Create an anchor chart with the shapes' names, drawings, and a real-life example of each, such as an ice cream cone. Encourage the students to reference the anchor chart during the lesson.
- Encourage the students to explore the shapes by touching them and explaining what they feel using nonverbal language such as gesturing or drawing and verbal language such as naming or describing the shape.

Students at this age have learned simple shapes such as triangles, circles, and rectangles. Here, students focus on recognizing and reproducing shapes based on specific attributes such as number of sides. In later grades, students will build on this understanding when computing area and perimeter.

Perform the **Try It!** activity on the next page.

Talk About It

- Have students use the Mini Relational GeoSolids® to compare the triangular prism and the square pyramid. **Ask:** *What are some ways we can compare these two shapes?* Guide students in noting that three-dimensional shapes can be compared by the number of faces, edges, and vertices. **Ask:** *How are these two shapes the same? How are they different?* Students should note that both have triangular faces, but the square pyramid only has 1 square face.

- Have the students use the Mini Relational GeoSolids® to practice sorting the shapes based on different criteria (for example, shapes with triangular faces and shapes without triangular faces).

Solve It

With students, reread the problem. **Ask:** *What is the name of the shape of the box? What is the name of the shape Kenji made?* Have students describe the attributes used to identify each shape.

More Ideas

For other ways to teach about shapes—

- Pair the students to play a guessing game. Have the first student describe one of the Mini Relational GeoSolids® shapes. Then, have the other student try to guess the shape described. Allow the student to ask questions to clarify, such as "Does it have a square base?"

- Have the students select a Mini Relational GeoSolids® shape. Then, ask the student to use the Geoboard to recreate one of the faces on the solid.

- For more practice, use Lesson 2 student page 54.

Try it Activity

15 minutes — Pairs

Here is a problem about shapes.

Kenji has a box. It has 6 flat faces and 8 vertices. The edges are all the same length. He traces one of the faces so he knows how large the bottom of the box is. The shape he traces has 4 sides that are all the same length. What are the names of the shape of the box and the shape he traced?

Introduce the problem. Then have students do the activity to solve the problem.

Distribute Geoboards and Mini Relational GeoSolids® to pairs.

1 **Say:** *Think about the shape that Kenji drew. What do you know about the shape?* Guide students to identify it as a two-dimensional shape with four sides that are the same length. **Say:** *Use your Geoboard to show the shape that Kenji drew. What shape did you make?*

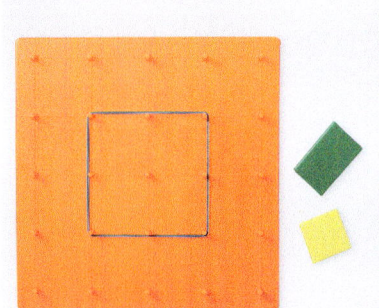

2 **Say:** *Look at the Mini Relational GeoSolids®. If you traced one of the faces, which would give you a square?* Guide students in identifying the cube and square prism. Students may place the different faces of the Mini Relational GeoSolids® on the geoboard to compare the number and length of edges/sides.

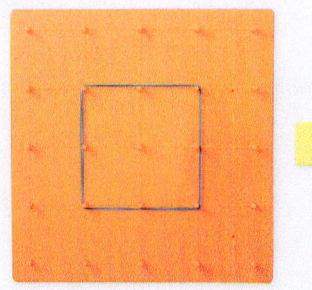

3 **Say:** *What do we know about the shape of Kenji's box that would help you identify the correct shape?* Students should note that the edges are all the same length. Have students compare the square prism and the cube to note that the cube has edges that are all the same length.

⚠ Look Out!

Watch out for students who are not able to differentiate between a rectangular prism and a cube. Have students count the number of faces, edges, and vertices to see they are the same. Then have students compare the lengths of the sides to see that all the edges of a cube are the same length, but a rectangular prism can have different edge lengths.

🔍 Formative Assessment

Have students try the following problem.

Lena picks up a GeoSolid® that has only triangles for all its faces. What shape did Lena pick up?

Lesson 3 Partition Rectangles

Objective
Partition rectangles into rows and columns.

Materials
- Color Tiles (20 per student)
- paper (1 sheet per student)
- markers (1 per student)

EL Support
- Review vocabulary: long, wide.
- Point out the terms *row* and *column*. Compare the everyday meaning of these words to the mathematical meaning.
- Provide students time to model a problem, write an answer, and pair up to share their answers.

Previously, students identified shapes, including rectangles. In this lesson, they apply the knowledge of rectangles to explore how square tiles can cover rectangles. This will help students in developing the skills needed for understanding multiplication and the concept of fractions.

Perform the **Try It!** activity on the next page.

Talk About It

Discuss the **Try It!** activity.

- **Ask:** *Will the answer change if there are 3 columns and 5 rows?*
- **Ask:** *Why is it important to make sure there are no gaps between the tiles?*
- **Ask:** *What if Sidney decides the box is too small or too large? How many squares would there be if she added a row? Added a column? Subtracted a row? Subtracted a column?*

Solve It

With students, reread the problem. **Ask:** *How many tiles does Sidney use to cover the top of the box?* Have students explain their solution using the Color Tiles.

More Ideas

For other ways to teach partitioning rectangles—

- Have students use a Geoboard and bands to make rectangles of 2 rows and 3 columns, 3 rows and 4 columns, and 4 rows and 4 columns. Tell them to partition each figure into squares to identify how many squares make each figure.

- Give students a sheet of 1-inch grid paper (Multi-Lesson BLM 1, page 86) and 12 tiles. Ask them to arrange all the tiles into rows and columns four different ways. Tell them to draw around the tiles after making each rectangle with the 12 tiles.

- For more practice, use Lesson 3 student page 55.

Try it Activity

20 minutes — Pairs

Here is a problem about partitioning rectangles.

Sidney is covering the top of a rectangular box with square tiles. The box top is 5 tiles long and 3 tiles wide. How many tiles will he need to cover the top of the box?

Introduce the problem. Then have students do the activity to solve the problem. Distribute Color Tiles, paper, and marker to students.

1
Have students choose 5 tiles. **Say:** *Line up 5 tiles on a sheet of paper in a row, making sure that the edges touch.* Discuss the importance of not leaving gaps. **Say:** *Draw a line along the top of the tiles.*

2
Say: *Now use 3 tiles to make a column.* Make sure the top tile of the column is in the same position as the first tile of the row they already made. **Say:** *Draw a line on the left side of the tiles.*

3
Say: *Now draw the other sides of the rectangle.* If students need guidance, have them use a ruler. **Say:** *Fill your rectangle with rows and columns of tiles.* **Ask:** *How many rows are in the rectangle?* **Ask:** *How many columns are in the rectangle?* **Ask:** *How many tiles fill your rectangle?*

⚠ Look Out!

Watch for students who don't line up the tiles next to one another. Make sure the students put the tiles in rows and columns with the edges touching so that there are no gaps between the tiles.

🔍 Formative Assessment

Have students try the following problem.

Aspen covers the top of a music box with 3 rows and 4 columns of tiles. How many tiles does Aspen use?

Lesson 4: Solve Problems by Partitioning Rectangles

Objective
Solve problems by partitioning a rectangle into rows and columns.

Materials
- Color Tiles (30 per student)
- paper (1 sheet per student)
- pencils (1 per student

EL Support
- Review vocabulary: rows, columns.
- Write the following sentence frame to be used during the **Try It!** When using tiles to make rows and columns, I notice _____.
- Allow time for students to discuss the process of placing tiles in rows and columns before determining the answer to the problem.

Students at this stage have placed Color Tiles in rows and columns and counted the total number of tiles. In this lesson, they apply the knowledge of covering rectangular spaces with tiles to identify the total number of tiles using addition.

Perform the **Try It!** activity on the next page.

Talk About It

Discuss the **Try It!** activity.

- **Ask:** *Why do we add 6 four times?*
- **Ask:** *Why is it important that the squares are equal in size?*
- **Ask:** *What would be the same and different if there were 6 rows with 4 bars in each row?*

Solve It

With students, reread the problem. **Ask:** *How many square bars are in the pan?* Have students explain their solution using the Color Tiles.

More Ideas

For other ways to teach solving problems by partitioning rectangles—

- Have students work in pairs. Give each pair a Geoboard and bands. Have one student choose the number of rows from 2–4 and the other student choose the number of columns from 2–4. Then have the students make the rectangle on the Geoboard and write an addition sentence for the rectangle. Have each pair repeat the activity for several different numbers of rows and columns.

- Give students a sheet of 1-inch grid paper (Multi-Lesson BLM 1, page 86), markers, scissors, and a number cube labeled 1–6. Tell them to roll the number cube once for the number of rows and again for the number of columns. Have students color a rectangle using the two numbers rolled. Then have them cut out the rectangle and write an addition number sentence for the rectangle.

- For more practice, use Lesson 4 student page 56.

Try it Activity

20 minutes — Pairs

Here is a problem about solving problems by partitioning rectangles.

Morgan is cutting a pan of bars into equal squares. There are 4 rows with 6 bars in each row. How many square bars are in the pan?

Introduce the problem. Then have students do the activity to solve the problem. Distribute Color Tiles, paper, and pencils to students.

1 **Say:** Let's use tiles to represent the pan of bars. **Ask:** How many rows of tiles are there? **Ask:** How many columns of tiles are there?

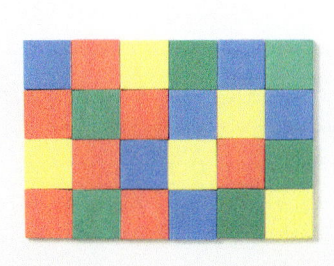

2 **Ask:** How many tiles are in the top row? **Say:** Write the number of tiles next to the top row. Watch that students place a 6 next to the top row. **Say:** Now count the tiles in each row and write the number of tiles next to each row.

3 **Say:** Now we will write an addition number sentence to find out how many bars are in the pan. Guide students to write a horizontal addition equation to show the number of tiles in each row. **Ask:** How many bars are in the pan?

$$6 + 6 + 6 + 6 = 24$$

⚠ Look Out!

Make sure students do not add the number of rows to the number of columns. Remind students that the number of tiles in each row is the same. So, the numbers added are also the same.

🔍 Formative Assessment

Have students try the following problem.

A rectangle is covered with 9 rows and 3 columns of tiles. How many tiles are there in all?

Geometry

Lesson 5 Partition Rectangles into Fair Shares

Objective
Partition a rectangle into rows and columns and describe shares using the words halves, thirds, half of, and a third of.

Materials
- Geoboard (1 per pair)
- paper (1 sheet per pair)

EL Support
- Review vocabulary: halves, thirds, fourths.
- Write the words *half,* and *third,* on the board, along with visual representations of the number of partitions they represent.
- Clarify the difference between ordinal words and fraction words by giving real-world examples such as *third in line* and *one-third of an apple.*

Students at this age have learned basic shapes. They have experience solving problems by partitioning rectangles into rows and columns. Here, students will expand this knowledge by partitioning rectangles into halves and thirds. Partitioning shapes will build a foundation for future work with fractions by helping students to understand and visualize equal parts.

Perform the **Try It!** activity on the next page.

Talk About It

Discuss the **Try It!** activity.

- Have students point to the partitioned shape they drew. **Ask:** *How many thirds?* Students should recognize that there are 3 thirds.
- Have the students create a rectangle that is 3 grid squares wide and 2 grid squares high. **Ask:** *How can you partition this shape into two equal parts?* Explain that each partitioned part is called a half. **Ask:** *How many halves are on the rectangle?*
- Ask the students to brainstorm examples of when they might partition a rectangular shape into halves or thirds. Students may mention such things as folding paper in thirds to fit in an envelope or cutting some fabric into halves to make a pillow.

Solve It

With students, reread the problem. **Ask:** *How many equal parts did you divide the garden into? How else can you name those parts?* Have students explain their reasoning using the Geoboard.

More Ideas

For other ways to teach partitioning—

- Use Color Tiles to model a rectangular flatbread that can be shared equally by two and three people. Explain that each piece is an equal portion of the rectangle. Have students identify the number of parts and name them as halves and thirds.
- Give the students a piece of paper shaped into a rectangle. Allow them to fold the paper into two equal shapes. Compare the different ways to partition the rectangle into halves.
- For more practice, use Lesson 5 student page 57.

Try it Activity

Here is a problem about partitioning.

Three friends are planting a garden. The garden plot is shaped like a rectangle. How can each friend have an equal amount of garden space to plant? How can we describe each share of the garden?

15 minutes — Pairs

Introduce the problem. Then have students do the activity to solve the problem.

Distribute Geoboards and paper to students.

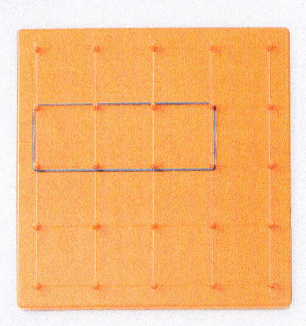

1
Say: *Let's make a model of the garden plot. What shape is the garden? What do you know about the sides of this shape?* Guide the students to create a rectangle that is 3 grid squares wide.

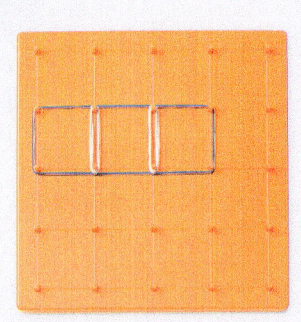

2
Ask: *How can the friends divide the garden plot into 3 equal parts?* Have students work with a partner to partition the rectangle into thirds. Allow the students to share their partitioned rectangles. **Ask:** *How do you know you have enough parts for each friend? How do you know the parts are equal?* Explain that when you divide a shape into equal parts, you partition the shape.

3
Say: *Let's draw the shape we made on our Geoboards.* Have students draw the rectangle on a separate sheet of paper. **Ask:** *How many equal spaces are there in the garden?* Explain that when you partition a shape into 3 equal pieces, each piece is called a third. Have the students point to each square with a partner and say, "a third."

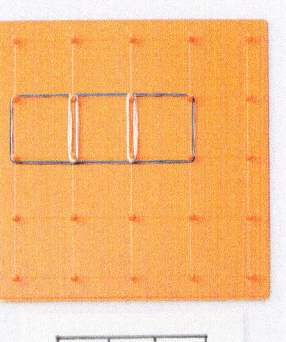

⚠ Look Out!

Watch for students who partition the rectangle into 3 rows and 3 columns instead of just 3 rows or 3 columns. Show students a rectangle partitioned in this way, and have them count the equal shares to see that there are 9 instead of 3.

🔍 Formative Assessment

Have students try the following problem.

Make a square on your Geoboard.

Partition the shape into 2 equal rows. What is each part called?

Lesson 6 Partition Circles

Objective
Partition circles into two, three, or four equal shares and describe the whole as two halves, three thirds, or four fourths.

Materials
- Rainbow Fraction® Circles (1 set per pair)
- paper (1 sheet per pair)

EL Support
- Review vocabulary: partition, halves, thirds, fourths. Point out the root word *part*. Explain that *partitioning* is dividing a shape into equal *parts*.

- Write the word *half* on the board. Explain that to make the plural of half, you change *-lf* to *-lves*. Point out similar plurals, such as *calf/calves* and *wolf/wolves*. Have students practice using *half* and *halves* in a sentence. For example, *I have two halves. I give one half to Josie.*

- Create visual displays to show halves, thirds, and fourths including an example and the term in singular and plural.

Students at this age have learned basic shapes. They have experience solving problems by partitioning shapes into rows and columns. Here, students will expand this knowledge by partitioning circles into halves, thirds, and fourths. Partitioning shapes will build a foundation for future work with fractions by helping students to understand and visualize equal parts.

Perform the **Try It!** activity on the next page.

Talk About It

Discuss the **Try It!** activity.

- Have students point to the partitioned shape they drew. **Ask:** *How many fourths did you make in the circle?* Students should identify there are 4 fourths.

- Ask students to draw a circle. **Ask:** *How can you partition this circle into halves? How can you describe the number of halves in the whole circle?* Students should describe the circle as having 2 halves. **Ask:** *How would you describe the whole circle if it were divided into thirds?* Students should describe the circle as having 3 thirds.

- Ask the students to brainstorm examples of when they might partition a circular object into halves, thirds, or fourths. Students may mention such things as cutting a quesadilla into pieces or making a craft.

Solve It

With students, reread the problem. **Ask:** *How many equal parts did you divide the cornbread into? How else can you name those parts?* Have students explain their reasoning using the Rainbow Fraction® Circles.

More Ideas

For other ways to teach partitioning—

- Have the students turn a Geoboard to the geocircle side. Challenge the students to partition the shape into halves, thirds, and fourths and name the sections for each as one-half, one-third, and one-fourth.

- Give the students paper plates. Allow them to cut the plates into halves. Have the students tell how many parts make up the whole: two halves. Repeat for thirds and fourths.

- For more practice, use Lesson 6 student page 58.

Try it Activity

15 minutes — Pairs

Here is a problem about partitioning circles.

Eddie made a round pan of cornbread. There are four people in the family. How can each family member have an equal share of the cornbread? How would you describe each share of the cornbread?

Introduce the problem. Then have students do the activity to solve the problem.

Distribute Rainbow Fraction® Circles and paper to students.

1 **Say:** *Let's make a model of the cornbread. Which shape can you use to represent it?* Guide students to select the whole unit fraction circle to represent the entire pan of cornbread.

2 **Ask:** *How can Eddie divide the cornbread into even parts for the family?* Allow the students to work with a partner to partition the circle into fourths by showing the 4 fractional pieces. **Ask:** *How do you know you have enough parts for each family member? How can you show that the parts are equal?* Remind the students that when you divide a shape into equal parts, you partition the shape.

3 **Say:** *Let's draw the shape we made with our Rainbow Fraction® Circles.* Allow the students time to draw the circle on a separate sheet of paper. **Ask:** *How many pieces of cornbread should they cut?* Explain that when you partition a shape into 4 equal pieces, each piece is called a fourth. Have the students point to each piece with a partner and say, "a fourth." Then have the students tell how many fourths there are all together.

Look Out!

Watch the students as they partition their circles and ensure that they divide their circle using segments that start at the center and not parallel lines through the circle to divide it into unequal parts. Have students return to the fractional circle pieces and stack them on top of each other to make sure they are equal. Then have the students cut out the parts they created and try to stack them to show that they are unequal parts.

Formative Assessment

Have students try the following problem.

Partition a circle into 3 equal parts. What is each part called?

Measurement

In second grade, students build upon their nonstandard measurement experiences by measuring in standard units. They use customary (inches and feet) and metric (centimeters and meters) units to measure lengths of objects by selecting appropriate tools. They select an attribute to be measured, choose an appropriate unit of measurement, and determine the number of units.

In second grade, students also measure an object using two units of different lengths (e.g., a desk measured in inches and in feet). Doing so helps students realize that the unit used and the attribute being measured are both important. They estimate lengths using inches, feet, centimeters, and meters, which helps them become more familiar with unit sizes. Students make connections between number lines and rulers. They use length to solve addition and subtraction word problems and create number lines with equally spaced points corresponding to whole numbers to solve problems to 100.

Students extend skip-counting by fives to tell and write time from analog and digital clocks to the nearest five minutes. They will also tell time to the nearest minute. Students solve word problems involving quarters, dimes, nickels, and pennies. They also use measurement data to solve word problems.

Measurement specifies that students should—

- Measure and estimate lengths in standard units.
- Relate addition and subtraction to length.
- Work with time and money.

The following hands-on activities provide students necessary experiences measuring, working with time and money, and solving word problems involving length. Using concrete materials prior to predicting, estimating, comparing, and solving problems enables students to acquire foundational concepts and gives them the confidence necessary to solve more difficult problems.

Measurement

Contents

Lesson 1 Estimate Lengths..................................22
   Objective: Estimate lengths using customary and metric units.
   Manipulative: Color Tiles, Cuisenaire® Rods

Lesson 2 Different Size Units..............................24
   Objective: Measure the length of an object twice and describe how the two measurements relate to the size of the unit chosen.
   Manipulative: Inchworms™, Inchworms™ and Centibugs™ Ruler

Lesson 3 Select and Use Measurement Tools.....................26
   Objective: Select and use appropriate tools to measure the length of an object.
   Manipulative: Inchworms™, Inchworms™ and Centibugs™ Ruler

Lesson 4 Measure and Compare Lengths........................28
   Objective: Measure and compare the lengths of two objects.
   Manipulative: Inchworms™

Lesson 5 Whole Numbers as Lengths on a Number Line............30
   Objective: Represent whole-number sums and differences as lengths on a number line.
   Manipulative: Cuisenaire® Rods

Lesson 6 Tell Time to 5 Minutes.............................32
   Objective: Tell and write time from analog and digital clocks to the nearest 5 minutes.
   Manipulative: NumberLine Clock™

Lesson 7 Tell Time to the Minute............................34
   Objective: Tell and write time from analog and digital clocks to the nearest minute.
   Manipulative: NumberLine Clock™

Lesson 8 Solve Coin Problems...............................36
   Objective: Solve word problems involving money.
   Manipulative: Coin Tiles and Hundred Board

Lesson 1 Estimate Lengths

Objective
Estimate lengths using customary and metric units.

Materials
- Color Tiles (15 per pair)
- Cuisenaire® Rods (30 white rods per pair)
- Measurement Recording Sheet 1 (Lesson 1, page 59, 1 per pair)

EL Support
- Review vocabulary: inches, centimeters, customary, metric, estimate.
- Explain that an estimate is different from a precise measurement. Have students give examples to explain when they might use an estimate using the following sentence frame. I might estimate _____ because _____.
- Encourage students to think of other situations when estimating length is useful, such as when you are trying to place a book in a shelf or need to decide how much wrapping paper you may need to wrap a present.

Students at this age have sorted objects by length, such as from shortest to longest. They also have experienced estimating lengths using nonstandard units such as blocks or paper clips. Here, students estimate lengths of objects using customary and metric units (inches and centimeters). This experience will build the foundation for students to solve problems using more precise measurements.

Perform the **Try It!** activity on the next page.

Talk About It

Discuss the **Try It!** activity.

- **Say:** *In this lesson, we used a customary measurement unit called inches.* **Ask:** *What are other customary measures of length?* Guide students in discussing feet. Show them a ruler and have them discuss when they might use feet and inches. **Ask:** *What is something that is about 1 foot long?*
- **Say:** *We also used a metric measure called centimeters.* **Ask:** *What are other metric measures of length?* Guide students in discussing meters. Show them a meterstick and have them discuss when they might use meters and centimeters. **Ask:** *What is something that is about 1 meter long?*
- **Ask:** *When is a time you have estimated something?*

Solve It

With the students, reread the problem. **Ask:** *How can you estimate the length of the straw in inches and centimeters? Do you think the bridge can be built by the straws you measured?* Have students explain using the Color Tiles and white Cuisenaire® Rods.

More Ideas

For other ways to teach estimating—

- Give students visual aids to estimate the length of an inch and a centimeter such as the width of two fingers for an inch and the width of the little finger for a centimeter.
- Set up stations of various classroom objects for students to estimate such as a book, a sheet of paper, a paper clip, and a pair of scissors. Then have the students use Inchworms™ to check their estimates.
- For more practice, use Lesson 1 student page 60.

Try it Activity

15 minutes | Pairs

Here is a problem about estimating lengths.

Mr. Tran's class is building bridges out of straws. The bridge cannot be taller than 10 inches or 25 centimeters. How can you estimate how many inches and centimeters a straw is in length?

Introduce the problem. Then have students do the activity to solve the problem.

Distribute 15 Color Tiles, 30 white Cuisenaire® Rods, and a Measurement Recording Sheet (Lesson 1, Page 59) to each pair.

1
Ask: *What does it mean to estimate? What helps you make a good estimate?* Review the customary and metric measurement systems. Show the students a Color Tile. Explain that 1 tile is 1 inch long. **Ask:** *About how many inches long is the straw?* Have the students estimate the length of the straw in inches and record it on the recording sheet.

2
Say: *Line up one Color Tile with the bottom of the straw. Add tiles in a row until they are at least as long as the straw.* Inform students that it is fine if the tiles are a little longer than the straw. Have the students write the length of the straw to the nearest inch on their recording sheets.

3
Ask: *What is another unit of measurement we could use to measure the length of the straw?* Show the students a white Cuisenaire® Rod. Explain that 1 white rod is 1 centimeter long. **Ask:** *About how many centimeters long is the straw?* Have the students estimate the length of the straw in centimeters and record it on the sheet.

4
Say: *Put one white rod at the bottom of the straw.* **Ask:** *How close do you think your estimate is?* Have the students measure the length of the straw using the blocks and then write the length to the nearest centimeter on their recording sheets.

⚠️ Look Out!

Watch for students who are not lining up the tiles or rods properly. They may leave gaps or not align with the end. Model for students the correct way to place the Color Tiles and Cuisenaire® Rods so that they have accurate measurements.

🔍 Formative Assessment

Have students try the following problem.

Estimate the length in inches and centimeters.

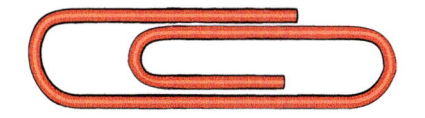

23

Lesson 2 Different Size Units

Objective
Measure the length of an object twice and describe how the two measurements relate to the size of the unit chosen.

Materials
- Inchworms™ (12 per pair)
- Inchworms™ and Centibugs™ Ruler (1 per pair)
- pencils (1 per student)
- Measurement Recording Sheet 2 (Lesson 2, page 61, 1 per pair)

EL Support
- Review vocabulary: inches, centimeters, customary, metric.
- EL students may be more familiar with the metric measurement system. Create a T-chart to compare metric and customary measures such as inch/centimeter and yard/meter. Show examples of each measure.

Students at this age have sorted objects by length such as shortest to longest. They also have experience with estimating lengths using nonstandard units, such as blocks or paper clips, and standard units, such as inches and centimeters. Here, students find exact lengths of objects using customary and metric units (inches and centimeters). This experience will build the foundation for students to solve complex problems using more precise measurements in later grades.

Perform the **Try It!** activity on the next page.

Talk About It

Discuss the **Try It!** activity.

- **Ask:** *What did you notice about the difference between the two measurements? Why do you think they were different?* Have students compare a centimeter and an inch. Guide students in understanding that since a centimeter is shorter than an inch, it takes more centimeters than inches to measure the same object.

- **Ask:** *How do you think a foot and a meter compare?* Have students compare a foot ruler and a meterstick. Guide the students in understanding that since a foot is shorter than a meter, it takes more feet than meters to measure the same object.

Solve It

With the students, reread the problem. **Ask:** *What was the measure of the water bottle in inches and centimeters? Will it fit in both the lunch box and the backpack?*

More Ideas

For other ways to teach measurement—

- Have students cut a strip from a 1-inch and a 1-centimeter grid paper and compare the two. Point out that although the strips are the same length, there are more centimeter boxes because the centimeter measure is smaller.

- Have students measure classroom materials using white Cuisenaire® Rods for measuring in centimeters and Color Tiles for measuring in inches.

- For more practice, use Lesson 2 student page 62.

Try it Activity

⏱ 15 minutes 👥 Pairs

Here is a problem about measuring.

Maria needs to check whether her new water bottle will fit in her lunch box and in her backpack. Her lunch box is 12 inches in length, while her backpack 26 centimeters in length. Measure the water bottle in inches and centimeters to see if it will fit in both the lunch box and the backpack.

Introduce the problem. Then have students do the activity to solve the problem.

Distribute 12 Inchworms™, an Inchworms™ and Centibugs™ Ruler, pencils, and a Measurement Recording Sheet 2 (Lesson 2, page 61) to each pair.

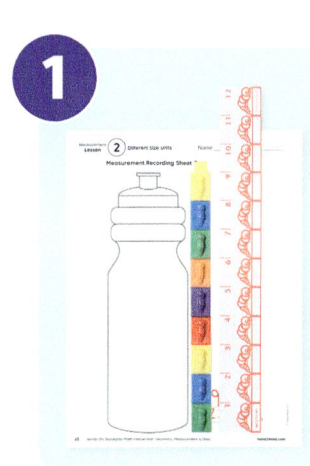

1
Ask: *What customary unit could we use to measure the height of the water bottle?* Have students line up a train of Inchworms™ next to the water bottle. Then have the students place the Inchworms™ Ruler next to the Inchworms™ to measure the height of the water bottle. Have the students record the length on the recording sheet.

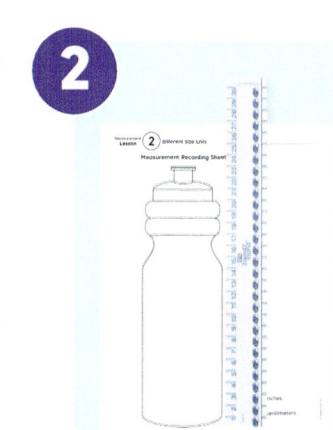

2
Ask: *What metric unit could we use to measure the height of the water bottle?* Have students place the Centibugs™ Ruler next to the water bottle to measure the length. Then have the students record the length on the recording sheet.

3
Ask: *How do the two measurements compare? Why is the number different for the measurement in inches and the measurement in centimeters?*

⚠ Look Out!

Watch out for students who think that a larger number of centimeters means a greater length for the water bottle. Remind students that even though the bottle is being measured using different units, the length remains the same.

🔍 Formative Assessment

Have students try the following problem.

Measure a cracker in inches and centimeters. Why are the two measurements different?

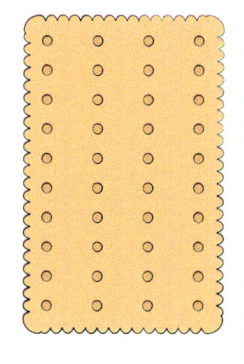

Lesson 3: Select and Use Measurement Tools

Objective
Select and use appropriate tools to measure the length of an object.

Materials
- Inchworms™ (12 per group)
- Inchworms™ and Centibugs™ Ruler (3 per group)
- pencils (1 per student)
- Measurement Recording Sheet 3 (Lesson 3, page 63, 1 per group)

EL Support
- Review vocabulary: ruler, yardstick, meterstick, measuring tape.
- Point out that *ruler*, *yard*, and *tape* are multi-meaning words. Discuss the meanings of each word, and explain the math meaning of each word. For instance, discuss that the word *ruler* can mean a person who leads a country or the instrument that is used to measure the length of objects. However, these words are used as words that relate to measurement in the context of this lesson.

Students have learned to estimate and measure lengths using nonstandard units such as blocks or paper clips and standard measurements such as inches and centimeters. Here, students will explore how to select the proper measurement tool to solve a measurement problem. This experience will build the foundation for students to solve problems using more precise measurements.

Perform the **Try It!** activity on the next page.

Talk About It

Discuss the **Try It!** activity.

- Discuss with students the measuring tools they used. **Ask:** *When would it be better to use an Inchworm™ than an Inchworms™ Ruler?*
- Show students a yardstick. **Ask:** *Would you use a yardstick to measure the door? Why?* Have students compare the ruler and the yardstick.
- Then show students a measuring tape. **Ask:** *How is this tool different from the yardstick and meterstick? When might you use this tool to measure?* Have students give examples of when a measuring tape is more efficient such as measuring how far a softball is thrown or how long a room is.

Solve It

With students, reread the problem. **Ask:** *Which measuring tool would you choose to measure the door? Why would you choose it? Which measuring tool would you use to measure the sticker? Why would you choose it?*

More Ideas

For other ways to teach measurement—

- Have students measure classroom materials such as crayons and pencils using Color Tiles and the Inchworms™ Ruler. Have them choose between the two and decide if they should measure crayons in inches or in feet.
- Brainstorm a list of lengths they could measure in the classroom and what tool they could use to measure each length. Discuss the advantages of each tool for that measurement (e.g., for measuring the length of the room with a measuring tape, the students wouldn't have to move the measuring tool over and over).
- For more practice, use Lesson 3 student page 64.

Try it Activity

15 minutes | **Groups of 6**

Here is a problem about measuring.

Mr. Nguyen's class wants to decorate their classroom door with stickers across its width. They need to measure the width of the door and the length of the stickers. Which tool should they use to easily measure each object?

Introduce the problem. Then have students do the activity to solve the problem.

Distribute 12 Inchworms™, Inchworms™ and Centibugs™ Rulers, pencils, and Measurement Recording Sheet 3 (Lesson 3, page 63) to each group. *Note: This* **Try It!** *assumes that classroom doors are of the standard width of 36 inches.*

1
Say: *Let's first measure the width of the classroom door.* Have one group try to line up Inchworms™ across the door to measure it. Have them mark the point on the door at which 12 Inchworms™ end and have another group start over from that point. Have groups continue to do this until they measure the entire door. Have students record their measurement in inches on Measurement Recording Sheet 3 (Lesson 3, page 63).

2
Ask: *Shall we try to measure the door again, but this time with the Inchworms™ Ruler?* Have one group of students use the ruler to measure the width of the door. Have them mark the point on the door at which one ruler ends end, and have another group start over from that point. Have them record the measurement in feet on the recording sheet.

3
Say: *Now, let's measure the length of the stickers.* Have students place the Inchworms™ Ruler under the sticker to measure the length of the stickers. Then have them use Inchworms™ to measure the sticker's length. **Ask:** *Do you think we should measure the stickers in feet or inches? Why?* Have students record their measurement in inches on Measurement Recording Sheet 3 (Lesson 3, page 63).

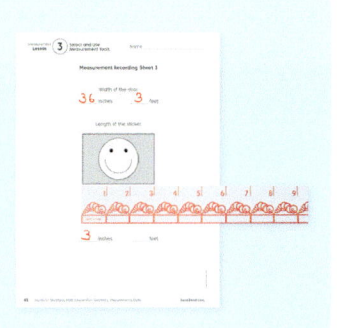

⚠ Look Out!

Watch out for students who leave gaps between Inchworms™ while measuring. Ensure that they are lined up properly. Look out for students who choose to measure longer objects with Inchworms™ and shorter objects with the ruler. Explain to them that they can estimate the size of objects first based on whether the object is closer to the length of an Inchworm™ (an inch) or a ruler (a foot).

🔍 Formative Assessment

Have students try the following problem.

Asa wants to measure the length of his desk. Which measurement tool should he use? Explain your reasoning.

Lesson 4 Measure and Compare Lengths

Objective
Measure and compare the length of two objects.

Materials
- Inchworms™ (12 per pair)
- pencils (1 per student)
- Measurement Recording Sheet 4 (Lesson 4, page 65, 1 per pair)

EL Support
- Review vocabulary: compare, inch.
- Discuss how to compare lengths. Use phrases such as "the worm is longer than the bug" and "the worm is 3 inches longer than the bug."

Students have learned to estimate and measure lengths using nonstandard units such as blocks or paper clips and standard measurements such as inches and centimeters. Here, students will explore how to compare the lengths of two objects using simple measurement tools. This will prepare students for measuring objects and creating graphs using the measurements as data.

Perform the **Try It!** activity on the next page.

Talk About It

Discuss the **Try It!** activity.

- **Ask:** *How did you compare the length of the two leaves?* Students should recognize that they can visually compare the two leaves as well as use the two measurements.
- **Ask:** *How would you compare the two leaves using centimeters?* Students should note that they could use the same procedure but with a different measuring tool.

Solve It

With the students, reread the problem. **Ask:** *Which leaf is longer? How much longer is it? How did you find out?* Have students explain their reasoning using the Inchworms™.

More Ideas

For other ways to teach measurement—

- Have students trace around classroom objects, such as a crayon or a pencil, on grid paper. Then have them use a Centibugs™ Ruler to measure the lengths of the objects and write the measurement next to each picture. Repeat using an Inchworms™ Ruler and compare the measurements.
- Have pairs measure the length of their feet using Color Tiles and then compare the measurements.
- For more practice, use Lesson 4 student page 66.

Try it Activity

15 minutes — Pairs

Here is a problem about comparing measurements.

Mrs. Garcia's class compares the measurement of two leaves during science class. Which is longer? How much longer is it?

Introduce the problem. Then have students do the activity to solve the problem.

Distribute 12 Inchworms™, pencils, and Measurement Recording Sheet 4 (Lesson 4, page 65) to each pair.

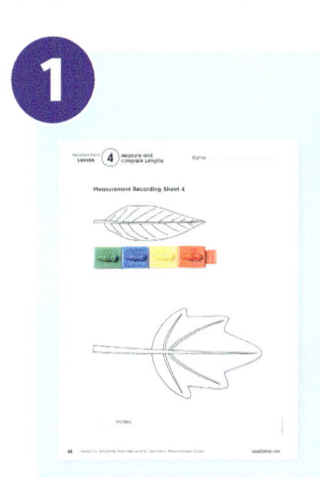

1

Ask: *How can you find the length of the top leaf?* Have students place the Inchworms™ on the recording sheet to measure the length of the first leaf. Then record the length.

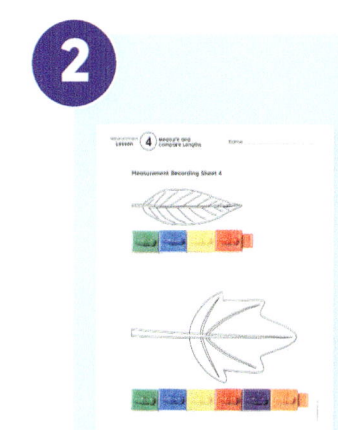

2

Ask: *What other information do we need if we want to compare the lengths of the two leaves?* Have students place the Inchworms™ on the recording sheet to measure the length of the second leaf. Then record the length.

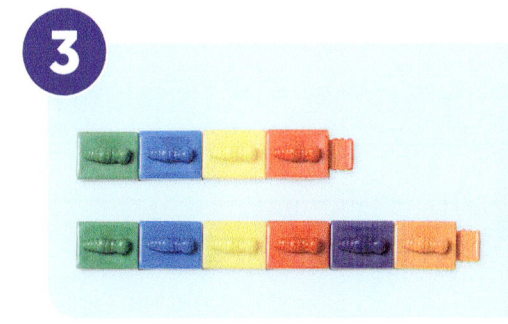

3

Have the students move the two groups of Inchworms™ to compare the lengths.
Ask: *How can you describe the difference between the two leaves?*

Look Out!

Watch for students who are not aligning the Inchworms™ with the edge of the drawing. Discuss why it is important to align the ruler with the start of the object. You may want to draw a grid line down from the drawing to help the students know how to align the worms.

Formative Assessment

Have students try the following problem.

Measure the length of two of your crayons and find out which one is shorter.

Lesson 5: Whole Numbers as Lengths on a Number Line

Objective
Represent whole-number sums and differences as lengths on a number line.

Materials
- Cuisenaire® Rods (1 set per pair)
- 1-cm Number Lines (Lesson 5, Page 67; 1 per pair)
- pencils (1 per student)

EL Support
- Review vocabulary: centimeter. Remind students that "centimeter" is abbreviated as "cm."
- Write the following sentence frame to be used during the **Try It!** activity. A number line can help me measure lengths by _____.

Students have learned estimating and measuring objects to the nearest unit using nonstandard measures such as cubes, as well as standard measurements such as inches and centimeters. In this lesson, students use Cuisenaire® Rods to show lengths on a number line. This will help them as they use number lines for other operations.

Perform the **Try It!** activity on the next page.

Talk About It

Discuss the **Try It!** activity.

- **Say:** *On this number line, each space between tick marks is the length of 1 white rod. The light green rod covers 3 spaces, and the black rod covers 7 spaces.*

- Write the number sentence 1 + 1 + 3 + 7 = _____ on the board. Have students correlate their actions from each step to the numbers in the sentence.

- **Ask:** *How does the number sentence show the length of the train?*

- **Ask:** *How is measuring an object using Cuisenaire® Rods and a number line like measuring an object using a ruler?*

Solve It

With students, reread the problem. Have students draw the rods on the number line and write number sentences that represent the value of the train and its length. **Ask:** *What is the length of Mikah's glasses?*

More Ideas

For other ways to teach about representing whole numbers as lengths—

- Have students work in pairs. Give each pair a copy of 1-inch Number Lines (Lesson 5, Page 68) and a handful of Color Tiles. Have one partner choose a number of Color Tiles of a single color and place them in a row on a 1-inch number line, starting at 0. Then have the other partner place a number of tiles of another color next to them on the number line. Have students work together to find the total length.

- Give students a copy of 1-cm Number Lines (Lesson 5, page 67) and a set of Cuisenaire® Rods. Call out a number of centimeters, and have students place 2 or more rods on a number line to make that exact length. As an additional challenge, call out more conditions, such as "use exactly 3 rods" or "no white rods."

- For more practice, use Lesson 5 student page 69.

Try it Activity

25 minutes — **Pairs**

Here is a problem about representing whole numbers as lengths on a number line.

Mikah got a new pair of glasses and he uses a train of Cuisenaire® Rods to measure the length of his glasses. He uses 2 white rods, 1 light green rod, and 1 black rod. At what number on the number line does the train end? What is the length of the train?

Introduce the problem. Then have students do the activity to solve the problem.

Distribute Cuisenaire® Rods, 1-cm Number Lines (Lesson 5, page 67), and pencils to pairs of students.

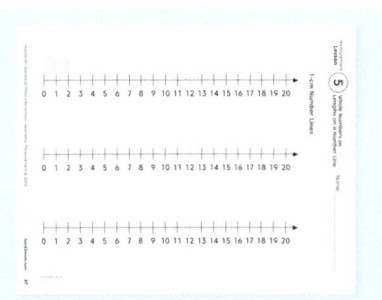

1

Say: *Let's start with the 2 white rods.* Have students place the 2 white rods in a train on the number line. Guide students to start at zero and build the train to the right. **Ask:** *At which number does the second white rod end? What is the length of the train?*

2

Say: *Now let's add the light green rod to the train.* Direct students to place the green rod at the end of the white rods without a gap. **Ask:** *Where does the train end now? What is the length of the train?* Write the number sentence 2 + 3 = 5 on the board. **Ask:** *How does this number sentence show the length of the rods?*

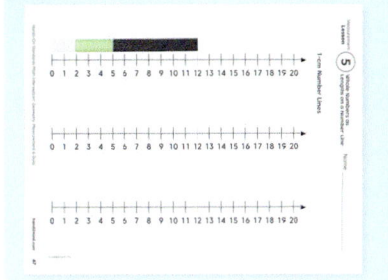

3

Say: *Add the black rod to the train.* Direct students to place the black rod at the end of the light green rod without a gap. **Ask:** *What is the length of the black rod? Where does the train end now?* **Say:** *You can use a number sentence to show the new action, too.* Write 5 + 7 = 12.

⚠ Look Out!

Watch for students who have difficulty remembering the values of the rods. You may wish to provide the students with a list of the rods and their measures. For example, a blue Cuisenaire® Rod is 9 cm long.

🔍 Formative Assessment

Have students try the following problem.

Mikah measures the length of his glasses case. He uses 1 yellow, 1 purple, and 3 red rods. How long is the glasses case on the number line?

Measurement

31

Lesson 6 — Tell Time to 5 Minutes

Objective
Tell and write time from analog and digital clocks to the nearest 5 minutes.

Materials
- NumberLine Clock™ (1 per pair)
- Telling Time Recording Sheet (Multi-Lesson BLM 2, page 87, 1 per pair)

EL Support
- Review vocabulary: minute, hour, and hand.
- Point out that *hand* is a multi-meaning word. Discuss the meaning of *hand* in the context of telling time. Have students practice using the word in different contexts, e.g. Can you hand me an apple? There are two hands on the clock. I have a ring on my right hand.
- Point out that a *digital* clock shows the time using *digits*, or numbers.

Previously, students have learned to tell time to the hour. In this lesson, students will practice telling time to 5-minute intervals. This will build the foundation for students to tell time to the minute.

Perform the **Try It!** activity on the next page.

Talk About It
Discuss the **Try It!** activity.

- **Ask:** *How do you know where to place the hour hand? If the hour is between the 4 and 5, what does that mean?*
- **Ask:** *If the minute hand is pointing to the red 8, how many minutes after the hour is it? How do you know?*
- **Ask:** *If the time is 8:30, how could you find which red number the minute hand is pointing to? If the time is 15 minutes after the hour, which red number is the minute hand pointing to? What if the time is 45 minutes after the hour?*

Solve It
With students, reread the problem Then have students draw clocks to show 11:50 and 1:15, and write the times in digital format.

More Ideas
For other ways to teach telling time to 5 minutes—

- Have the students take out the minute chain and the hour chain and unfold them with the number line side facing up. Have students work in pairs. One student should position the minute hand pointing to a number of minutes that ends in 0 or 5. Then the other student should tell which hour number the hour hand would be pointing to.
- Create a class schedule. Have the students show the times on the NumberLine Clock™ as well as write them in digital form.
- For more practice, use Lesson 6 student page 70.

Try it Activity

20 minutes — **Pairs**

Here is a problem about telling time to 5 minutes.

Mrs. Nguyen's class goes to lunch at 11:50 a.m. They finish lunch and recess at 1:15 p.m. The class has an analog clock, and Mrs. Nguyen has a digital watch. How will the class know when it is time to go to lunch and when to come back?

Introduce the problem. Then have students do the activity to solve the problem.

Distribute NumberLine Clocks™, Telling Time Recording Sheets (Multi-Lesson BLM 2, page 87), and pencils to students.

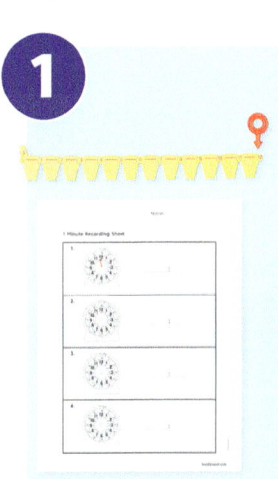

1

Say: *Lunch starts at 11:50 p.m. That time is between 11:00 and 12:00.* **Ask:** *Which time is it closer to?* Have students unfold the hour chain with the number line side facing up. Guide students to point the hour hand between the 11 and 12, closer to the 12. **Say:** *Place the hour chain back in the base and move the hour hand to the correct position. Draw the hour hand on Clock 1 on your recording sheet.*

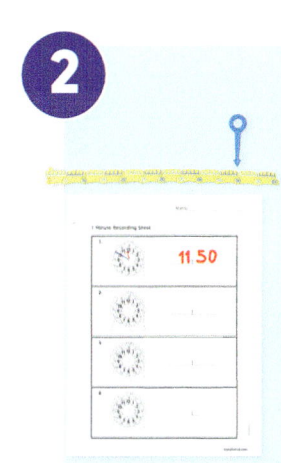

2

Say: *The small blue numbers on the NumberLine Clock™ show the minutes. There is a number for every 5 minutes.* Have students unfold the minute chain. *Let's move the minute hand as we count together by 5s.* Point to the small blue numbers on the clock as you count together by 5s to 50. **Say:** *Place the minute chain back in the base and move the minute hand to the correct position. Draw the minute hand and write the time in digital format on your recording sheet.*

3

Say: *The class finishes lunch and recess at 1:15.* **Ask:** *Where should the hour hand point to at this time?* Have students unfold the hour chain with the number line side facing up. Guide them to point the hour hand right past the 1. **Say:** *Place the hour chain back in the base and move the hour hand to the correct position. Draw the hour hand on Clock 2 on your recording sheet.*

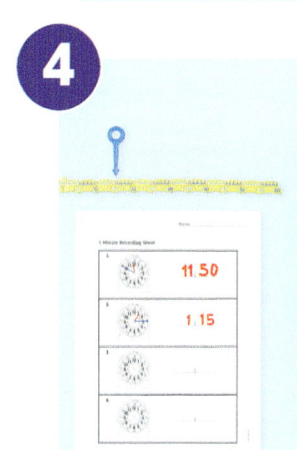

4

Say: *Let's see where the minute hand should point to.* Have students unfold the minute chain with the number line side facing up. Guide them to point the hour hand to 15. **Say:** *Place the minute chain back in the base and move the minute hand to the correct position. Draw the minute hand on Clock 2 on your recording sheet.* Direct students to write the time in digital format on the recording sheet.

⚠ Look Out!

Watch for students who always place the hour hand exactly on the hour number. Use the clock to show the students that unless it is exactly 3, the hour hand should be placed between the two hours. Give students times in digital format and have them indicate if the hour hand should be closer to one number or another.

🔍 Formative Assessment

Display a clock with the time shown below. Have students try the following problem.

Jonah's dad tells him to walk his dog at the time shown. What time should Jonah walk his dog?

Lesson 7 Tell Time to the Minute

Objective
Tell and write time from analog and digital clocks to the nearest minute.

Materials
- NumberLine Clock™
- Telling Time Recording Sheet (Multi-Lesson BLM 2, Page 87, 1 per pair)

EL Support
- Review vocabulary: hour, minute, and hands.
- Explain the use of the phrases "in a minute" and "give me 5 minutes." Students may have confusion as these phrases are often used loosely and don't represent an exact duration of time.
- Reduce the language load in the **Try It!** problem by simplifying the task to show and write the time.

Students have learned to tell time to the hour, half hour, and 5-minute intervals. In this lesson, they will deepen their understanding by telling time to the nearest minute. Understanding time will support the students later as they solve problems with time, including elapsed time.

Perform the **Try It!** activity on the next page.

Talk About It

Discuss the **Try It!** activity.

- Display a NumberLine Clock™. **Say:** *The blue numbers tell us the number of minutes. The hour starts over at 60, so 60 minutes is the same as 0.* Point to the 60 and then count by 5s with the students, pointing to each number as you go around the clock, until you get back to 60.

- **Ask:** *What strategy did you use to count the minutes?* Discuss counting by 5s and then adding individual minutes to count more efficiently. Some students may suggest starting at 15 and counting by 5s and 1s from there.

- **Ask:** *Where did you place the hour hand? Why wasn't it placed on the 8?*

Solve It

With students, reread the problem. Then, have students show the time on the NumberLine Clock™ that Bella shows on her digital clock.

More Ideas

For other ways to teach telling time to the minute—

- Students can create a clock to take home for practice using a small paper plate, brad, and construction paper hands. Students can fold the clock in half, and then fold in half again perpendicular to the first fold to locate 12, 3, 6, and 9 and to find the center of the plate. Then have them estimate the positions of the remaining hours.

- Have students work in pairs. One partner should write a time on the digital clock, and the other partner should show the same time on the NumberLine Clock™. Then have students switch roles and repeat.

- For more practice, use Lesson 7 student page 71.

Try it Activity

15 minutes   **Pairs**

Here is a problem about telling time to the minute.

Bella's digital clock shows the time as 8:21. Where would the hands point on an analog clock?

Introduce the problem. Then have students do the activity to solve the problem.

Distribute NumberLine Clocks™ and Telling Time Recording Sheets (Multi-Lesson BLM 2, Page 87) to pairs of students.

1

Say: *A digital clock uses numbers and a colon to show the time. The hour is written before the colon, and the number of minutes is written after the colon.* Guide students to write the time on their recording sheet in digital format.

2

Have students unfold the hour chain with the number line side facing up. **Ask:** *Where should you place the hour hand between 8:00 and 9:00? How do you know?* Students should indicate that 8:21 is almost halfway to 9:00. Guide the students to replace the hour chain, move the hour hand to the correct position, and draw the hour hand on the clock on the recording sheet.

3

Say: *Let's find the position of the minute hand.* Have students unfold the minute chain with the number line side facing up. Count together with students by 5s, pointing at the numbers as you count. **Say:** *5, 10, 15, 20.* **Ask:** *Why do we stop at 20?* Elicit that if you keep counting to the next number by 5s, you would go past 21. **Say:** *Now, count up from 20 to 21.*

4

Guide the students in inserting the chain into the base of the clock and moving the minute hand to the correct position. **Say:** *Draw the minute hand on the clock on the recording sheet.* Remind the students that the minute hand is longer than the hour hand.

⚠ Look Out!

Watch for students who struggle with understanding where to place the hour hand. Teach children the positions for the minute hand for 15, 30, 45, and 60 minutes on the clock so they can read the clock more quickly.

🔍 Formative Assessment

Have students try the following problem.

Nikki's alarm is set for 4:47 to remind her to clean her room before dinner. How can you show this on an analog clock?

Measurement

35

Lesson 8 Solve Coin Problems

Objective
Solve word problems involving money.

Materials
- Coin Tiles (1 set per pair)
- Hundred Board (1 per pair)
- Hundred Chart (Lesson 8, Page 72; 1 per pair)
- pencils (1 per student)
- crayons (1 set per pair)

EL Support
- Review vocabulary: dollar, bill, quarter, dime, nickel, penny.
- Create an anchor chart with the name, image, and value of each coin for the students' reference.
- Phrase some questions so that they can be answered with limited language such as yes/no, true/false, or hand signals.

Students have learned how to identify pennies, nickels, dimes, and quarters as well as their values. In this lesson, students will utilize this understanding to solve problems involving coins. This will support students later as they add, subtract, multiply, and divide with money.

Perform the **Try It!** activity on the next page.

Talk About It

Discuss the **Try It!** activity.

- Have students recap how they used the Coin Tiles to figure out which number they would have to count up from and which number they would have to count up to. **Ask:** *How did you know to start at 35? How did you know to count up to 50?*

- Encourage students to look at this problem as double-digit addition. **Ask:** *Do you think you could have solved this problem without the Coin Tiles?* Discuss ideas, and model the addition sentence 10 + 10 + 10 + 5 = 35.

- **Ask:** *How else can you find out how much more money Arianna needs?* Remind students that to find how much more you need, you have to find the difference between two amounts.

- **Ask:** *What coins can she add to have exactly 50 cents?*

Solve It

With students, reread the problem. Have students draw the coins Arianna starts with and write the total amount needed to buy the eraser. Then, have them draw and write how much more she needs to buy the eraser.

More Ideas

For other ways to teach problem solving with coins—

- Set up a small store for the students, or use images with price tags. Have the students work with a partner to pretend to buy things and make change.

- For students who need more support counting money, you may wish to place "coin dots" on a visual display. Each dot represents 5, so a quarter has 5 dots, a dime has 2 dots, and a nickel has 1 dot. The students can count by 5s by touching the dots to count the coins.

- For more practice, use Lesson 8 student page 73.

Try it Activity

35 minutes — **Pairs**

Here is a problem about pennies, nickels, dimes, and quarters.

Arianna wants to buy an eraser for 50 cents at the book fair. She has 3 dimes and 1 nickel. How much more money does she need to buy the eraser?

Introduce the problem. Then have students do the activity to solve the problem.

Distribute Coin Tiles, Hundred Boards, Hundred Charts (Lesson 8, page 72), pencils, and crayons to students.

1 **Say:** *First, let's put all the coins Arianna has on the Hundred Chart. Start with the dimes, and then add the nickel.* **Ask:** *How much money does Arianna have to spend?* Have students remove the tiles and circle 35 using a red crayon.

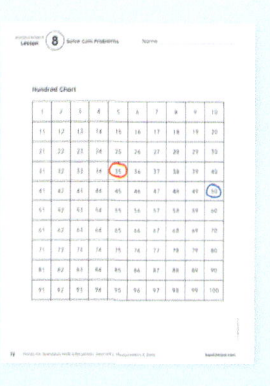

2 **Ask:** *How much does the eraser cost?* Have students circle the 50 on the Hundred Chart using a blue crayon.

3 **Say:** *To find out how much more Arianna needs, find the difference between the cost and how much she has. So, you can count up from 35 to 50 to find out how much more money Arianna needs.*

⚠ Look Out!

Watch for students who calculate the amount of money Arianna has instead of how much she still needs in order to buy the eraser. Remind students that Arianna does not have enough money, so they need to find the difference.

🔍 Formative Assessment

Have students try the following problem.

In the cafeteria, ice cream costs 65 cents. Pablo pays 75 cents. How much change should he get?

Measurement

Data

In second grade, students build on their skills of representing and interpreting data. Students have been introduced to bar graphs and pictographs. They continue their data skill development by exploring additional ways to present data, including picture graphs and line plots. Students continue developing data and problem-solving skills by working with different data displays.

Students use measurement data as they pose questions and collect, analyze, and represent data to interpret the results. They represent the lengths of several objects by making a line plot, where the horizontal scale is marked off in whole-number units. They draw picture graphs and bar graphs to represent data sets with up to four categories. They also solve problems involving addition, subtraction, and comparison, using information presented in line plots, bar graphs, and picture graphs.

Data specifies that students should—

Represent and interpret data.

The following hands-on activities provide students with additional experience measuring and representing data. Students deepen their problem-solving skills as they analyze and compare data represented in a graph. Their experiences with manipulatives will lead to greater understanding of how to select a data display that best shows the data. Manipulatives can provide a concrete representation of data sets. Physical manipulation of the manipulatives helps students explore the data and the organization of data in meaningful ways.

Data

Contents

Lesson 1 Line Plots Using Inches 40
    Objective: Measure and graph the length of four objects in inches using a line plot.
    Manipulative: Inchworms™

Lesson 2 Line Plots Using Centimeters 42
    Objective: Measure and graph the length of four objects in centimeters using a line plot. .
    Manipulative: Cuisenaire® Rods

Lesson 3 Solve Problems Using a Line Plot 44
    Objective: Solve problems using information presented in a line plot.
    Manipulative: Inchworms™ and Inchworm™ Ruler

Lesson 4 Picture Graphs. 46
    Objective: Draw a picture graph to represent a data set.
    Manipulative: Mini Relational GeoSolids®

Lesson 5 Bar Graphs. 48
    Objective: Draw a bar graph to represent a data set and use the information to solve problems.
    Manipulative: Coin Tiles, Cuisenaire® Rods

Lesson 6 Solve Problems Using Graphs. 50
    Objective: Solve problems using information presented in a bar graph.
    Manipulative: Cuisenaire® Rods

Lesson 1 Line Plots Using Inches

Objective
Measure and graph the length of four objects in inches using a line plot.

Materials
- Inchworms™ (10 per pair)
- Bookmark Line Plot Recording Sheet (Lesson 1, page 74, 1 per pair)

EL Support
- Review vocabulary: line, plot.
- Discuss the everyday meanings of the words *line* and *plot*. Have the students give examples of lines such as when the class lines up for lunch or a line on a road. Explain to students that a plot can be a piece of land or what happens in a story.
- Write the following sentence frame to be used during the **Try It!** I can find the length by _____.

Students at this age have learned how measure lengths. Here, students use that skill to measure the length of objects and plot these on a line plot. This will lay the foundation for students to create and plot other types of graphs.

Perform the **Try It!** activity on the next page.

Talk About It

Discuss the **Try It!** activity.

- **Ask:** *What does each X represent?* Students should understand that each X represents one object with that length.

- **Ask:** *What can you tell from comparing the height of the different columns?* Students should explain that the height of the plot above each measurement is the number of things with that length. By looking at the heights of the different columns, they can quickly compare the number of things with that height.

- **Ask:** *What are some other things you might place on a line plot?* Students may mention other measurements such as the lengths of nails in a tool kit or the number of glasses a restaurant has in each size.

Solve It

With the students, reread the problem. **Ask:** *What was the length of each bookmark? What did the line plot show?*

More Ideas

For other ways to teach measuring and creating a line plot—

- Have students use Color Tiles to measure classroom objects such as a crayon, a pencil, and paper. Then have students use that data to create a line plot.

- To create a line plot, give the students a sticky note with an X on it. Create a number line from 0 to 10 on the board. Have the students place their sticky note above the number that represents the number of siblings they have. Discuss the findings.

- For more practice, use Lesson 1 student page 75.

Try it Activity

Here is a problem about measuring and graphing.

20 minutes   Pairs

Tameika's class made bookmarks of different lengths. She wants to see how many bookmarks they made of each length. How can Tameika measure the items? How can she compare the measurements?

Introduce the problem. Then have students do the activity to solve the problem.

Distribute 10 Inchworms™ and the Bookmark Line Plot Recording Sheet (Lesson 1, page 74) to each pair.

1

Ask: *How can you find out how long each bookmark is? Where should you place the first Inchworm™?* Have students measure each bookmark and record the length on the blank below it.

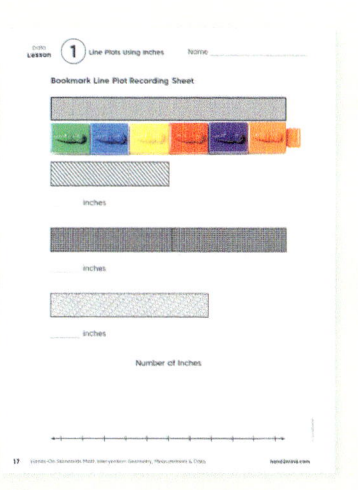

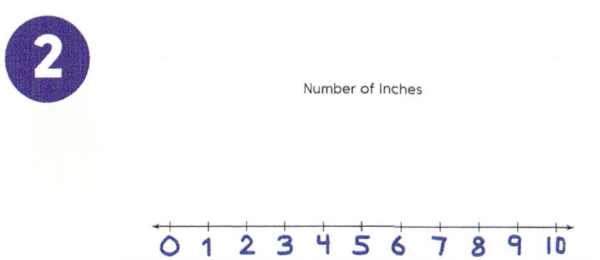

2

Say: *One way to compare the lengths of the bookmarks is to create a line plot. Look at the number line on the recording sheet. Think about the measurements you recorded.* **Ask:** *What number should we start with? What number should we write next?* Guide students to write an appropriate scale below the number line. Make sure they do not skip numbers that are not represented in the data.

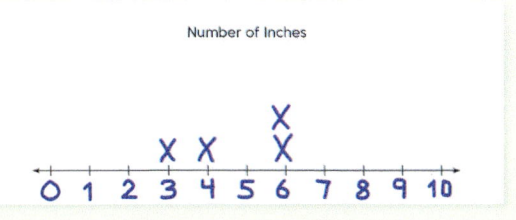

3

Ask: *How long was the first bookmark?* **Say:** *We can show the bookmark that measured 6 inches by placing an X above the 6.* Have students record all four lengths on the line plot. **Ask:** *Where did you place Xs? What did you do when there was more than 1 bookmark with the same length?*

⚠️ Look Out!

When asking questions about the number of bookmarks of each length, watch out for students who are using the length of the object and not the line plot. Remind the students that they are comparing the number of bookmarks of each length.

🔍 Formative Assessment

Have students try the following problem.

Kamar measures the length of the things in his pencil bag. His eraser is 2 inches long. His glue stick is 5 inches tall. His scissors are 7 inches long, and his pencil sharpener is 2 inches long. Make a line plot to show the data.

41

Lesson 2 Line Plots Using Centimeters

Objective
Measure and graph the length in centimeters of four objects.

Materials
- Cuisenaire® Rods (white rods, 25 per pair)
- Nail Line Plot Recording Sheet (Lesson 2, page 76, 1 per pair)

EL Support
- Review vocabulary: line plot.
- Explain that the word *plot* can be a noun or verb. Explain that it can be a verb when you are describing how to record the information, as in "plot the data," or a noun when you are describing the display, as in "draw a line plot."
- When asking questions about "most" when discussing line plots, clarify that you are looking at the frequency of the occurrence, not the longest length, as "most" can refer to both of these uses.

Students at this age have learned how to measure lengths in both inches and centimeters. Here, students use that skill to measure the length of objects in centimeters and plot these on a line plot. This will lay the foundation for students to create and plot other types of graphs. In later grades, students will use line plots to organize information to find the range and mode of a data set.

Perform the **Try It!** activity on the next page.

Talk About It

Discuss the **Try It!** activity.

- **Ask:** *How many nails measured 8 cm? Should the line plot include a mark for 8?* Guide students in understanding that it is important to include all the numbers on the number line even if they don't have a data point because it helps you to compare the measurements.

- Have the students compare the line plots from this lesson with the ones from the previous lesson. **Ask:** *How are they different? How are they the same?* Students should note that the line plots from the previous lesson showed the measurements in inches, while the ones in this lesson show them in centimeters. **Ask:** *How would the line plot change if you switched to measuring in inches?* Guide students to understand that since the size of the objects doesn't change, the frequency of each measurement would not change.

Solve It

With the students, reread the problem. **Ask:** *What was the length of each nail? What did the line plot show?*

More Ideas

For other ways to teach measurement and line plots—

- Have students use other Cuisenaire® Rods to measure objects in the classroom. Then create a line plot to compare the frequency of the different measurements.

- Students can use grid paper and stickers to create a line plot. Have the students gather measurement data such as how long their foot is and then place a sticker on the class chart to indicate their measurement. Then analyze the data as a class to look at the frequency of each measurement.

- For more practice, use Lesson 2 student page 77.

Try it Activity

20 minutes | Pairs

Here is a problem about measuring and graphing.

Malia is organizing her toolbox. She wants to see how many nails she has in each length. How can she measure the nails? How can she display the measurements?

Introduce the problem. Then have students do the activity to solve the problem.

Distribute 25 White Cuisenaire® Rods and the Nail Line Plot Recording Sheet (Lesson 2, page 76) to each pair.

1

Say: *How can you find out how long each nail is in centimeters? Where should you place the first Cuisenaire® Rod?* Have students measure each nail in centimeters and record the length on the blank below it.

2

Say: *A line plot helps us to see how many nails have each length. Think about the measurements you recorded.* **Ask:** *How can we label the number line?* Guide students to write an appropriate scale below the number line.

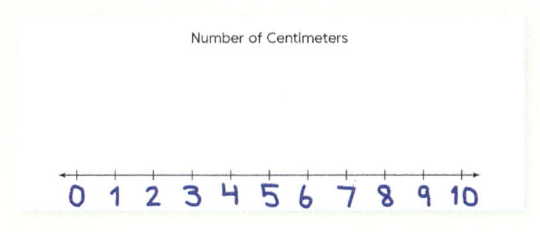

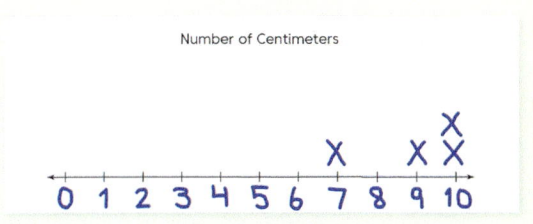

3

Ask: *How long was the first nail? We can show the nail that measured 7 centimeters by placing an X above the 7.* Have students record all four lengths on the line plot. **Ask:** *Where did you place Xs? What did you do when there was more than 1 nail with the same length?*

⚠ Look Out!

Watch out for students who do not line the rods up correctly to measure. Have students use a ruler to create a straight edge down from the start of the eraser. Have them place the first rod flush with the ruler and each of the remaining rods touching the one before it.

🔍 Formative Assessment

Have students try the following problem.

Bryce measures the length of some of his toys. He makes a list of their lengths.
Fire truck: 8 cm
Plane: 9 cm
Car: 6 cm
Van: 9 cm

Make a line plot to show the data.

Lesson 3 Solve Problems Using a Line Plot

Objective
Solve problems using information presented in a line plot.

Materials
- Inchworms™ (8 per pair)
- Inchworms™ Rulers (1 per pair)
- Line Plots Recording Sheet (Lesson 3, page 78, 1 per pair)
- crayons, erasers, and colored pencils (5 per pair)

EL Support
- Review vocabulary: graph, line plot.
- Remind students that groups of words work together in phrases. Review the phrases "is taller than" and "is shorter than." Have the students compare their heights using the phrases.
- Encourage students to use the graph as a visual aid when discussing their data. Students may point to different areas of the graph to explain their thinking.

Students at this age have learned how to measure lengths in both inches and centimeters. They have also learned how to create line plots. Here, students use their understanding of line plots to solve measurement problems. As students advance, they will build upon this understanding to create and analyze other types of graphs and displays.

Perform the **Try It!** activity on the next page.

Talk About It

Discuss the **Try It!** activity.

- Have students discuss how they measured the items. **Ask:** *How can you make sure your measurements are accurate?* Students may mention making sure the Inchworms™ are lined up end to end with no gaps.
- Allow time for students to share their line plots. **Ask:** *How does the line plot help you analyze the data? What does each column show? How many are shorter than 3 inches?*
- Have the students compare their line plots with another pair. **Ask:** *Why might your line plot be different from another group's line plot?*

Solve It

With students, reread the problem. Have students complete the line plot by adding a title. Then have students compare their line plots and write a sentence telling how many supplies are longer than 3 inches.

More Ideas

For other ways to teach line plots—

- Have students use white Cuisenaire® Rods to measure classroom objects in centimeters. Then have the students create a line plot of their data. Discuss the data with the students, such as how many objects were shown for each measurement.
- Have students measure the length of their feet to the nearest inch using a measuring tape. Have them create a tally chart with the lengths in inches. Mark these numbers on a line plot that shows the lengths for the whole class.
- For more practice, use Lesson 3 student page 79.

Try it Activity

20 minutes | Pairs

Here is a problem about creating line plots.

Mrs. Long's class is putting their art supplies in baskets. They want to know how many supplies are longer than 3 inches so they can choose the right basket. How can they use a line plot to show this information?

Introduce the problem. Then have students do the activity to solve the problem.

Distribute Inchworms™, Inchworms™ Rulers, and the Line Plots Recording Sheet (Lesson 3, page 78) to students.

1

Have each student choose five art supplies to measure, for example: erasers, colored pencils, or crayons. Ask students to use the Inchworms™ and Inchworms™ Ruler to measure the first item to the nearest inch. **Ask:** *How did you find the length? What did you do if the length was not exactly a whole number of inches?*

2

Direct students to place a tally mark for the correct length of the crayon on the tally chart. **Ask:** *What does the tally mark show?* Then have pairs measure and record tally marks for the remaining four items, placing the measurements on the tally chart.

3

Once the students have completed the tally chart, guide them to record the information on the line plot. **Ask:** *What is the first length on your tally chart?* Direct students to record an X above that number on the line plot. Then have students continue by adding an X above the length for each tally in their tally chart. Circulate to make sure that students are placing the X in the box above each number. **Ask:** *Where should you place the second X if there are two items with the same length?*

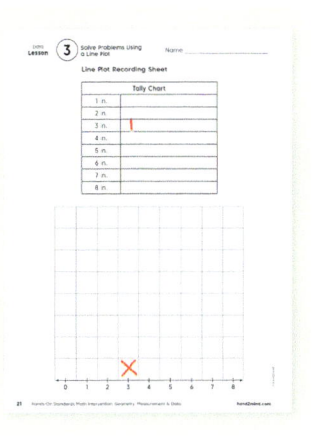

⚠ Look Out!

The **Try It!** problem asks students to count the objects that are more than 3 inches. Watch for students who include things that measure exactly 3 inches. Students may use an index card to cover the items that measure 3 inches and below.

🔍 Formative Assessment

Have students try the following problem.

Pedro is making a craft out of fabric. The line plot shows the lengths of fabric that Pedro cut to make the craft. How many lengths of fabric are less than 6 inches?

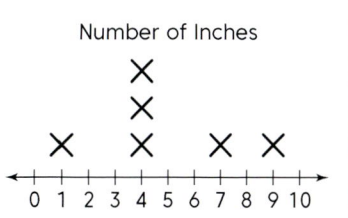

Number of Inches

Data

45

Lesson 4 Picture Graphs

Objective
Draw a picture graph to represent a data set.

Materials
- Mini Relational GeoSolids® (10 per pair)
- Picture Graphs Recording Sheet (Lesson 4, page 80, 1 per pair)
- crayons (1 each red, green, yellow, and blue per pair)

EL Support
- Review vocabulary: picture, graph. Show examples and discuss how a picture graph is different from other types of graphs.
- Review the names of the shapes: cone, cube, pyramid, cylinder. You may wish to create an anchor chart with the name and an image for students to reference during the lesson.
- Point out that the word *graph* is both a noun and a verb. Have the students use *graph* in both forms. For example, *I graph our food choices. I put an X on the graph.*

Students have learned how to find patterns and organize objects. They have used their pattern-recognition skills to organize data and create a line plot. In this lesson, students will organize and represent data using a picture graph. This will build an understanding for future work with analyzing and understanding bar graphs.

Perform the **Try It!** activity on the next page.

Talk About It

Discuss the **Try It!** activity.

- Have students look at their completed graphs. **Ask:** *How does this graph organize the data? What is another way you could organize the shapes? What would that graph look like?*
- **Say:** *Compare your graph to another group's graph.* **Ask:** *How are they similar? How are they different?*
- **Ask:** *What could be another title for the graph?*

Solve It

With students, reread the problem. Discuss how the graphs help show the data. **Ask:** *How could you use a picture graph to show how many solids of each color they have?*

More Ideas

For other ways to teach picture graphs—

- To help students see the connection between real objects and picture graphs, have 10 students draw a picture of their favorite fruit. Then ask students to stand in a different row for each fruit holding their pictures. As a class, have students count the number of students standing in each row. Then have the students put their picture on the floor as a placeholder and return to their seats. Discuss the graph that is left behind and how each picture represents the student's individual choice.
- Give students different numbers of attribute blocks. Have them arrange them in rows differentiated by shape and count the number of attribute blocks in each row. Have students create a picture graph to represent the number of shapes.
- For more practice, use Lesson 4 student page 81.

Try it Activity

20 minutes **Pairs**

Here is a problem about organizing a data set into a picture graph.

Mr. Garcia's class is having a shape scavenger hunt. They are looking for everyday objects shaped like cubes, cylinders, rectangular prisms, and cones. They keep track of how many they find. How can they use this information to make a picture graph of the items collected?

Introduce the problem. Then have students do the activity to solve the problem.

Distribute crayons; a collection of 10 cubes, cylinders, rectangular prisms, and cones from the Mini Relational GeoSolids®; and the Picture Graphs Recording Sheet (Lesson 4, page 80) to each pair of students.

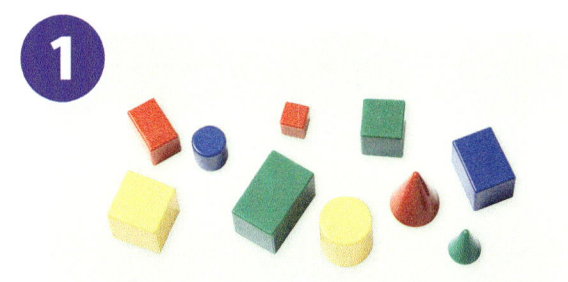

1
Give each group a random assortment of 10 Mini Relational GeoSolids®. **Ask:** *What solids of each shape do you see?* Have students sort the solids by shape.

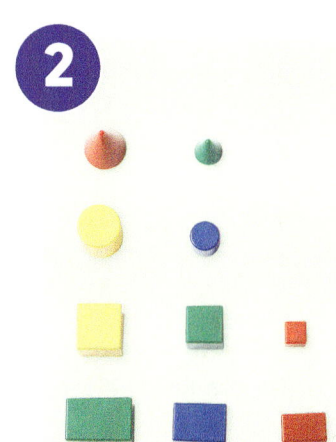

2
Say: *Let's put this information in a picture graph.* Have the students use the Mini Relational GeoSolids® to create a picture graph. **Ask:** *How could we organize the solids so that it is easy to compare the number of each?* Have students organize each type of solid in a row. **Ask:** *How can we space them so it is easier to compare the number of each?* Students should indicate that equal spacing helps you to compare the numbers.

3
Next, have students start making the picture graph by writing the shapes on the left side. **Say:** *Draw a cone for each cone you counted.* Guide students to draw the correct number of cones. Point out that the color of the cone does not have to match the cones shown. Discuss why you might want all the cones to be the same color on the graph. **Say:** *Now draw the other shapes using different colors.* Have students complete the title and label.

⚠ Look Out!

Watch for students who are not setting up their rows and columns correctly. Point out that just like in reading, a graph is usually read from left to right.

🔍 Formative Assessment

Have students try the following problem.

Create a picture graph for the data below.
3 cubes
2 cones
4 cylinders
1 pyramid

Data

47

Lesson 5 Bar Graphs

Objective
Draw a bar graph to represent a data set and use the information to solve problems.

Materials
- Coin Tiles (1 of each value per pair)
- Cuisenaire® Rods (1 set per pair)
- Bar Graph Recording Sheet (Lesson 5, page 82, 1 per pair)
- crayons (1 each purple, red, and green per pair)

EL Support
- Review vocabulary: bar graph.
- Explain that *bar* is a multi-meaning word. Show students examples of a bar on a bar graph and have them use the word in this context.
- Review the names of the coins. You may wish to provide the students with a reference page for their math notebook with the name, value, and a picture of each coin.

In previous lessons, students have learned how to create and analyze line plots and picture graphs. In this lesson, the students will create bar graphs. In later lessons, students will use bar graphs to solve problems.

Perform the **Try It!** activity on the next page.

Talk About It

Discuss the **Try It!** activity.

- **Ask:** *What do the lengths of the bars on the graph tell us about the number of coins for each type?*
- **Ask:** *Which coin did Liam have the most of? Which coin did he have the least of? How do you know?*
- **Ask:** *How would the graph change if Liam had 5 quarters?*

Solve It

With students, reread the problem. Have students compare the lengths of the rods representing the number of each coin and explain which coin Liam has the most of.

More Ideas

For other ways to teach drawing bar graphs—

- Students may practice making bar graphs by using UniLink® Cubes. For example, students may create a graph of their shoe colors. Place a pile of UniLink® Cubes and a piece of paper with color names on a desk. Have students place one cube each on the paper next to the color of their own shoes. Explain that each cube represents the shoe color of one student. Discuss how you could link the cubes to create a bar graph. Replace shoes with another item if all students wear the same color shoes as part of a school uniform.

- Have students collect data and use Color Tiles to create bar graphs. It might be easier for students to understand if they use tiles of all one color in each bar of the graph.

- For more practice, use Lesson 5 student page 83.

Try it Activity

20 minutes — **Pairs**

Here is a problem about creating bar graphs.

Liam has a handful of change and sorts it into 3 pennies, 4 nickels, 2 dimes, and 3 quarters. Make a bar graph of Liam's coins. Which coin does he have the most of?

Introduce the problem. Then have students do the activity to solve the problem.

Distribute crayons, Coin Tiles, Cuisenaire® Rods, and the Bar Graph Recording Sheet (Lesson 5, page 82) to students.

1
Distribute Coin Tiles to pairs. **Ask:** *What coins did Liam have in his pile?* Allow students to name the coins. **Ask:** *Which coin tile shows a penny?* Have students use Coin Tiles to represent the number of each type of coin.

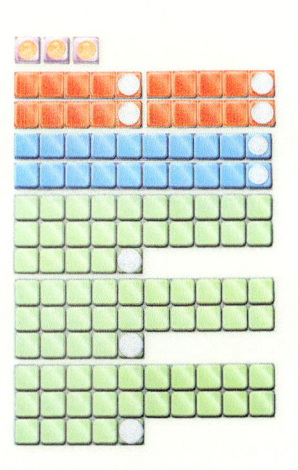

2
Ask: *How many pennies does Liam have? Which Cuisenaire® Rod has that length?* Have students use 3 white rods to find the correct length. Repeat for the number of nickels, dimes, and quarters. Guide students to use a green rod for the number of pennies and quarters, a purple rod for the number of nickels, and a red rod for the number of dimes.

3
Have students start making the bar graph by writing a title and the names of coins. Have them place the Cuisenaire® Rods in each column to represent the bars of the graph. Then have them use crayons to trace around the rods and color the bars. **Ask:** *How can we use the graph to find the number of coins Liam has the most of?*

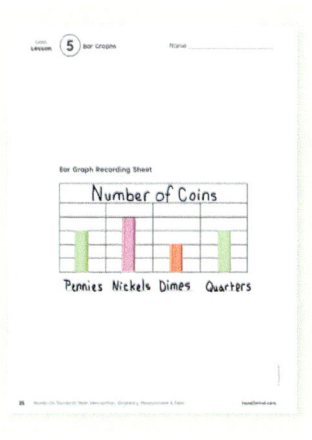

⚠️ Look Out!
Watch for students who do not use the correct length of rod. You may wish to have the students arrange a set of each color in order from shortest to longest to help them see the different lengths.

🔍 Formative Assessment
Have students try the following problem.

Mikhail's bank has 2 pennies, 3 nickels, 2 dimes, and 4 quarters. Make a bar graph showing the number of each type of coin.

Data

Lesson 6 Solve Problems Using Graphs

Objective
Solve problems using information presented in a bar graph.

Materials
- Cuisenaire® Rods (1 set per pair)
- Insect Recording Sheet (Lesson 6, page 84, 1 per pair)
- crayons (1 each blue, purple, green, and red per pair)

EL Support
- Review vocabulary: bar graph.
- Review the terms *put together, take apart*, and *compare*. Discuss the actions used and have the students model them using blocks or counters.
- Use hand gestures to clarify meaning. For example, when discussing *put together*, make a motion of moving two objects together, and for *take apart*, make a motion of moving an object away.

Students have learned how to create and analyze line plots, picture graphs, and bar graphs. In this lesson, students will expand this understanding by creating bar graphs to solve problems. In the future, this fluency with using data will support the students as they explore more complex types of data representation including different types of graphs.

Perform the **Try It!** activity on the next page.

Talk About It

Discuss the **Try It!** activity.

- **Ask:** *Which insect did Tiana see the most of? Which did she see the least of? How can you see that on the graph?*
- **Ask:** *How did you know how many butterflies she saw?*
- **Ask:** *How would the problem be different if you were asked how many dragonflies and butterflies she saw?*

Solve It

With students, reread the problem. Have students calculate the total length of the rods representing the butterflies and caterpillars and explain the answer.

More Ideas

For other ways to teach solving problems using bar graphs—

- Students can practice creating bar graphs using Color Tiles. Have each student name his or her favorite book. Once all students have shared their favorites, write the book names on sticky notes—one book name per sticky note. Then have the students place a Color Tile next to their favorite book to create a bar graph of the class favorites.
- Make a bar graph out of UniLink® Cubes. Create descriptors for students' lunch choices. Have the students add a cube to the linked train to indicate their choice. Discuss how the information could be used to communicate with the cafeteria staff.
- For more practice, use Lesson 6 student page 85.

50

Try it Activity

20 minutes | **Pairs**

Here is a problem about graphs.

While Tiana was working in her garden, she saw 9 beetles, 4 butterflies, 6 caterpillars, and 2 dragonflies. How can Tiana make a graph to show the number of each type of insect she saw? How many butterflies and caterpillars did she see in all?

Introduce the problem. Then have students do the activity to solve the problem.

Distribute Cuisenaire® Rods and the Insect Recording Sheet (Lesson 6, page 84) to students.

1
Say: *Use the Cuisenaire® Rods to represent the number of each type of insect.* Have students find the rods of lengths 9, 4, 6, and 2. **Say:** *Place the rods on the recording sheet to make a bar graph.* Guide students to label the columns and title the graph.

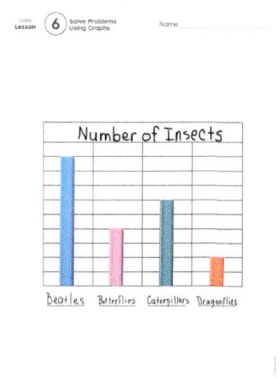

2
Have students trace the outlines of the rods on the graph and color them in. **Ask:** *How can we find the number of butterflies and caterpillars Tiana saw?* Guide students to place the rod showing the number of butterflies next to the rod showing the number of caterpillars.

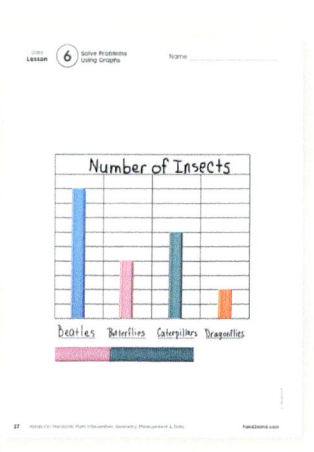

3
Ask: *How can we find the length of the two rods put together?* Have students find a rod with the same length as the two rods. **Ask:** *What is the length of the orange rod? How many butterflies and caterpillars did Tiana see?*

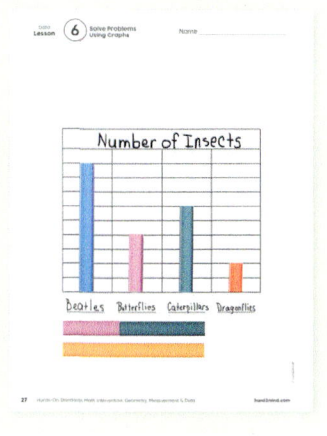

⚠ Look Out!

Watch out for students who try to add the number of rods instead of the lengths of the rods. Have the students place white rods above the rods to understand the value of each rod.

🔍 Formative Assessment

Have students try the following problem.

Display the bar graph that shows the number of each color flower Talia planted in her garden. How many more red flowers did she plant than purple flowers?

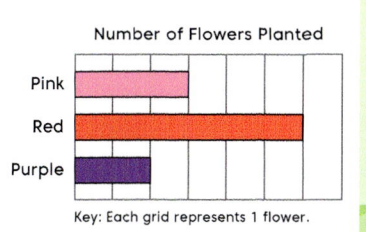

Number of Flowers Planted
Key: Each grid represents 1 flower.

Data

51

Geometry Lesson 1: Identify Shapes

Name _____

Dot Paper

Geometry **Lesson 1** Identify Shapes Name _____

Circle the name of the shape on the Geoboard.

1.

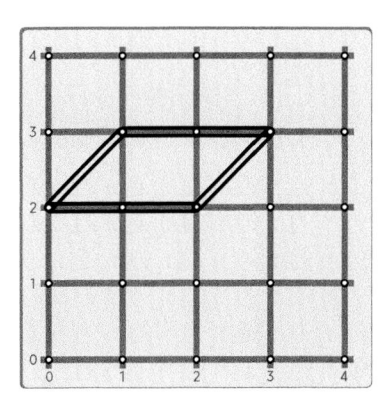

pentagon   quadrilateral

hexagon   triangle

2.

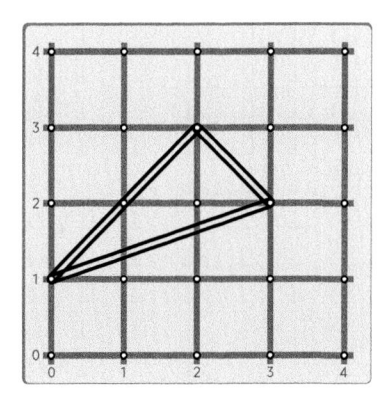

pentagon   quadrilateral

hexagon   triangle

3.

pentagon   quadrilateral

hexagon   triangle

4.

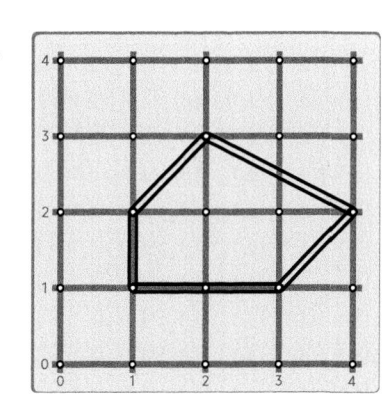

pentagon   quadrilateral

hexagon   triangle

Challenge! Tomas makes a shape on a Geoboard. He says that it is a quadrilateral. His friend says that it is a rectangle. Can they both be correct? Explain.

Hands-On Standards Math Intervention: Geometry, Measurement & Data   hand2mind.com   53

Geometry **Lesson 2** Recognize and Draw Shapes

Name _____

Circle the shape that matches the description.

1. This shape has 6 sides.

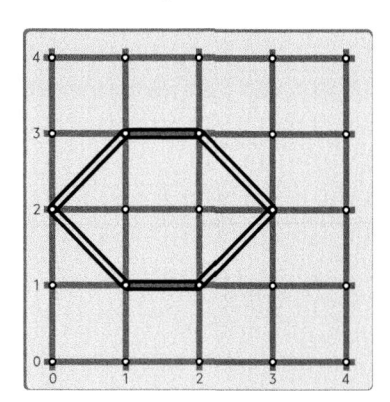

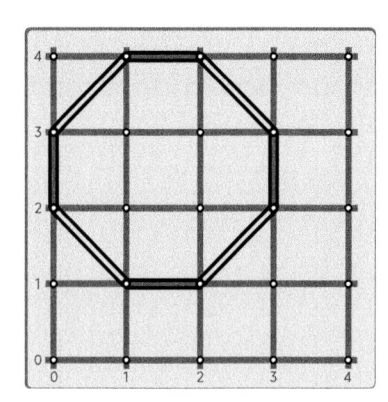

2. This shape has 6 faces. 4 faces are rectangles, and 2 faces are squares.

3. This shape has 3 sides.

    circle   square   triangle

4. This shape has 5 sides.

    rectangle   pentagon   triangle

Challenge! Name three real-life objects that are rectangular prisms.

54   Hands-On Standards Math Intervention: Geometry, Measurement & Data   hand2mind.com

Geometry **Lesson 3** Partition Rectangles   Name _____

Use Color Tiles to build the model. Write how many tiles are used.

1.

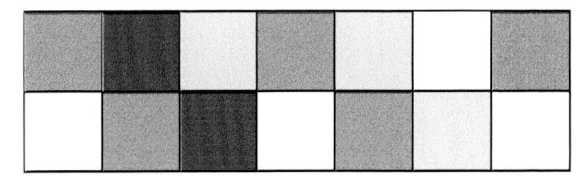

   _____ tiles

2.

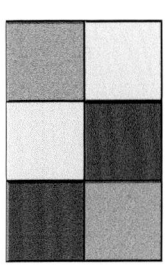

   _____ tiles

Use Color Tiles to build the model. Write how many tiles are used.

3. 5 rows and 2 columns   _____ tiles

4. 1 row and 4 columns   _____ tiles

5. 6 rows and 2 columns   _____ tiles

6. 3 rows and 5 columns   _____ tiles

Challenge! Austin uses rows and columns of square tiles to fill a rectangle. Each row has the same number of tiles, and each column has the same number of tiles. He uses exactly 18 tiles. How many rows and columns could Austin make with the tiles?

Geometry **Lesson** **4** Solve Problems by Partitioning Rectangles

Name _____

Use Color Tiles to build the model. Complete each number sentence to find how many tiles are used.

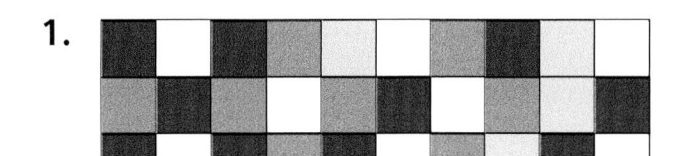

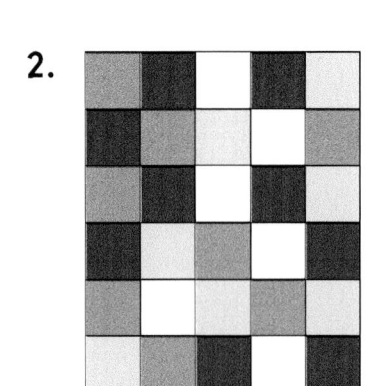

1.

_____ + _____ + _____ = _____ tiles

2.

_____ + _____ + _____ + _____ + _____ + _____ = _____ tiles

Use Color Tiles to build the model. Write a number sentence to show how many tiles are used.

3. 2 rows and 8 columns _____ = _____ tiles

4. 9 rows and 3 columns _____ = _____ tiles

5. 5 rows and 2 columns _____ = _____ tiles

6. 4 rows and 7 columns _____ = _____ tiles

Challenge! Fiona places square placemats on the table. She places the same number of placemats in each row. She uses exactly 24 placemats. How many rows and columns could Fiona make with the placemats? If she has 4 fewer placemats, how will the arrangements differ?

Geometry
Lesson 5 — Partition Rectangles into Fair Shares

Name _____

Make the shape on your Geoboard. Partition the shape and write the name of each part.

1. Make 2 equal parts.

   _____

2. Make 3 equal parts.

   _____

Circle the word that describes each part.

3.

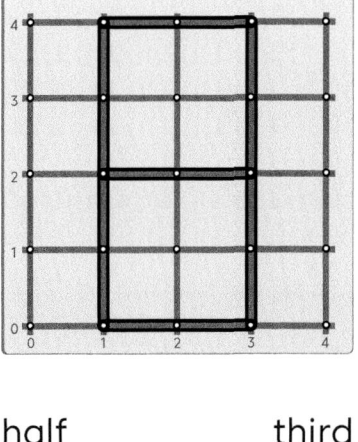

half   third

4.

half   third

Challenge! Ms. Rivera made a shape on a Geoboard. Keira partitioned the shape into 2 halves. Sasha partitioned the same shape into 3 thirds. Draw a possible shape with the different partitions.

Geometry **Lesson 6** Partition Circles   Name _____

Draw a box around the circle that is partitioned into the given number of parts.

1. 2 equal parts

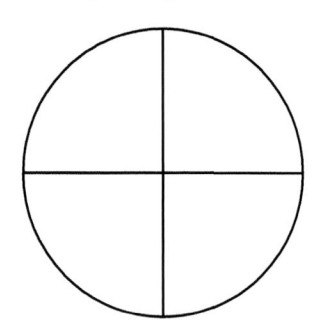

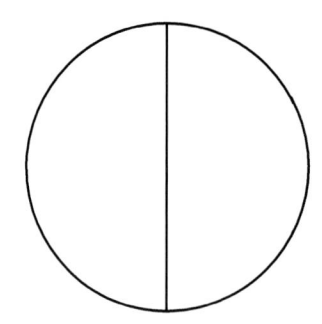

2. 3 equal parts

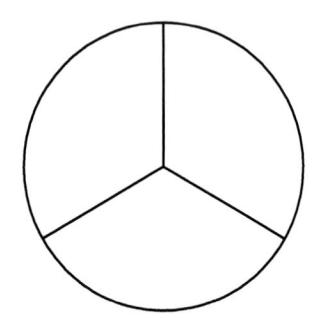

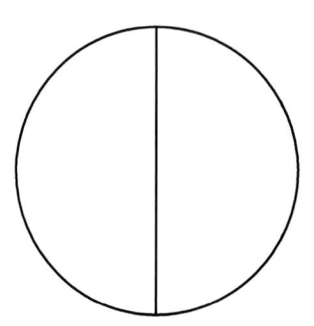

Partition the circle into the given number of parts using Rainbow Fraction® Circles.

3. 4 equal parts

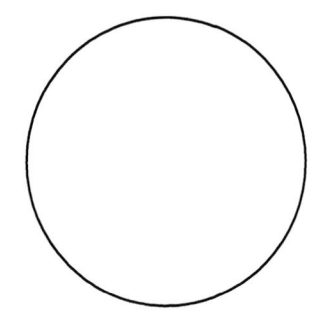

4. 2 equal parts

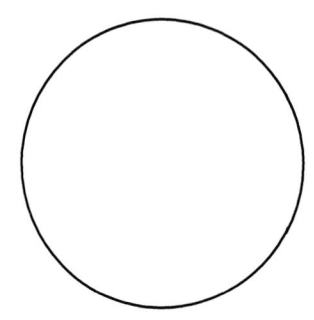

Challenge! How can you describe one part of a circle that is divided into 2 equal parts? How can you describe all the parts that make up the whole circle?

Hands-On Standards Math Intervention: Geometry, Measurement & Data   hand2mind.com

Measurement
Lesson 1 Estimate Lengths

Name _____

Measurement Recording Sheet 1

Estimate

_____ inches

_____ centimeters

Actual length

_____ inches

_____ centimeters

Hands-On Standards Math Intervention: Geometry, Measurement & Data

Measurement
Lesson **1** Estimate Lengths

Name _____

Estimate the length of each animal in inches and centimeters. Then measure them using Color Tiles and Cuisenaire® Rods and record the length.

1.

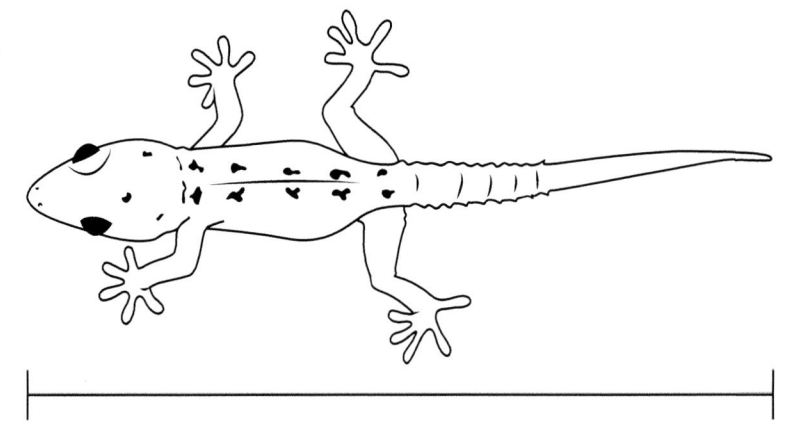

Estimate

_____ inches

_____ centimeters

Actual length

_____ inches

_____ centimeters

2.

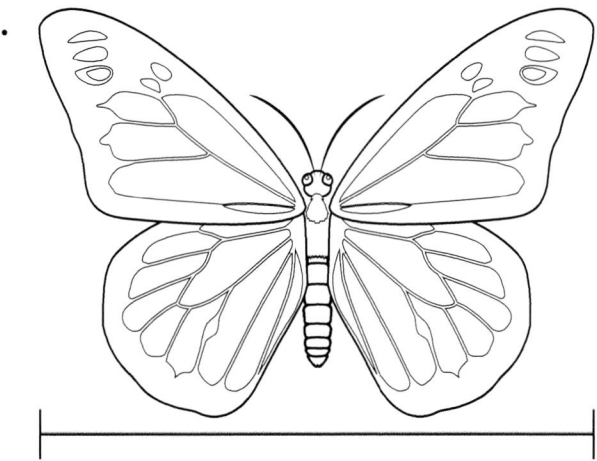

Estimate

_____ inches

_____ centimeters

Actual length

_____ inches

_____ centimeters

Challenge! Lena buys a new pencil. She estimates that it is 3 cm long. Do you think this is a good estimate? Explain why.

Measurement Lesson 2 — Different Size Units

Name _____

Measurement Recording Sheet 2

_____ inches

_____ centimeters

Hands-On Standards Math Intervention: Geometry, Measurement & Data

Measurement
Lesson 2 Different Size Units

Name _____

Measure the object in inches and centimeters.

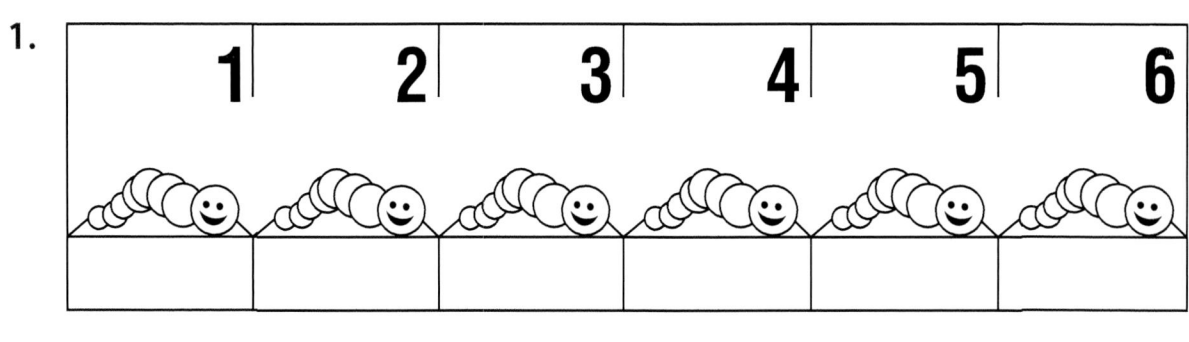

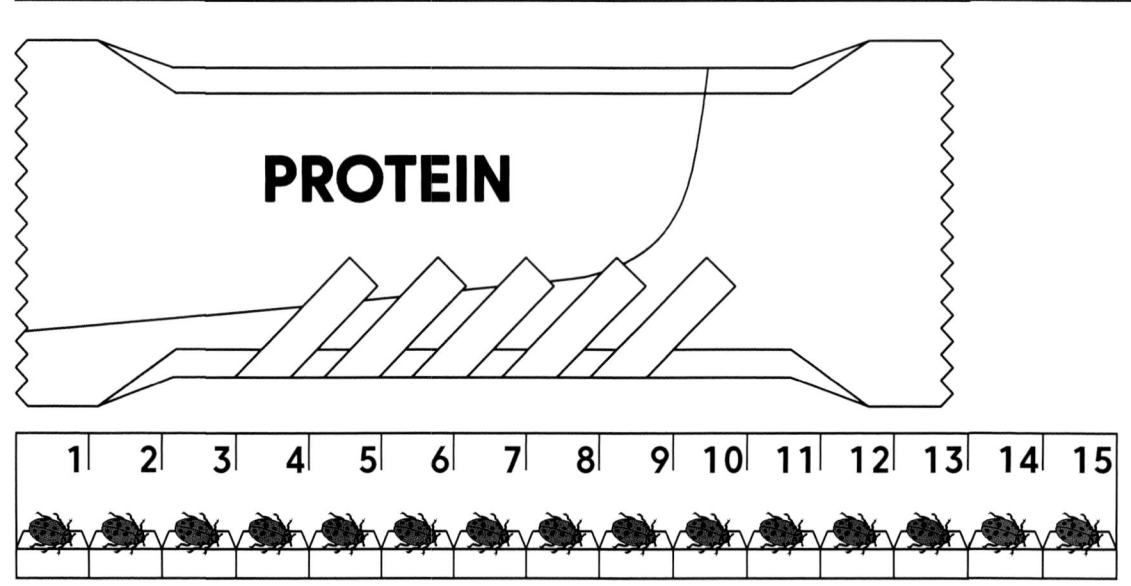

1.

_____ inches   _____ centimeters

Challenge! Abbey and Jacob measure a toy car using different units. Abbey's answer is 5 units, and Jacob's answer is 13 units. Who measured in centimeters, and who measured in inches? Explain how you know.

Measurement **Lesson** ③ Select and Use Measurement Tools

Name _____

Measurement Recording Sheet 3

Width of the door

_____ inches _____ feet

Length of the sticker

_____ inches _____ feet

Measurement
Lesson (**3**) Select and Use Measurement Tools

Name _____

Circle the better measurement tool. Then write the length.

1.

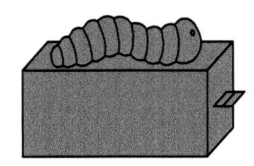

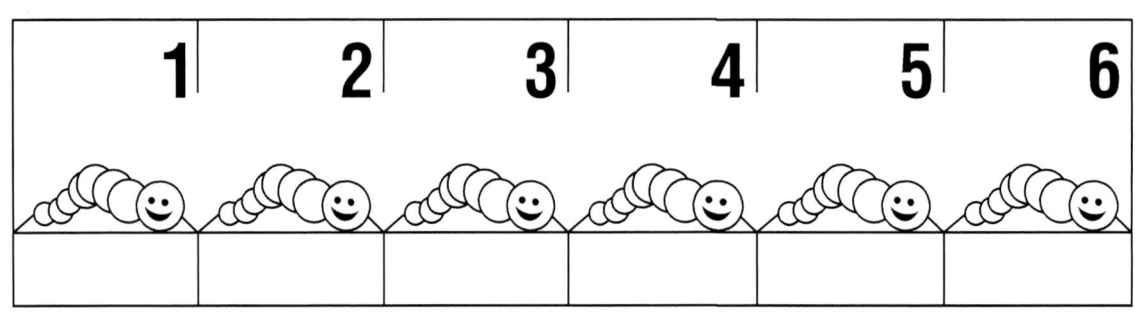

_____

Choose a measurement tool. Measure the object and write the length. Include units in your answer.

2.

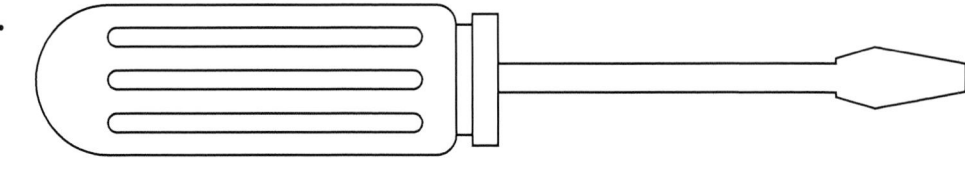

tool: _____   length: _____

Challenge! Sarika is making a macaroni necklace. She needs to measure the length of each piece of macaroni as well as the string. Which tool should she use for each?

64 Hands-On Standards Math Intervention: Geometry, Measurement & Data   hand2mind.com

Measurement Lesson (4) **Measure and Compare Lengths**

Name _____

Measurement Recording Sheet 4

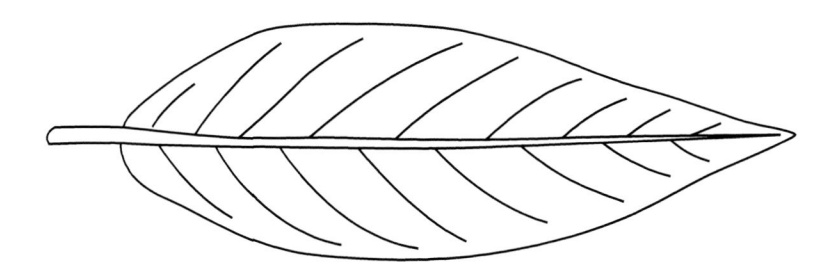

_____ inches

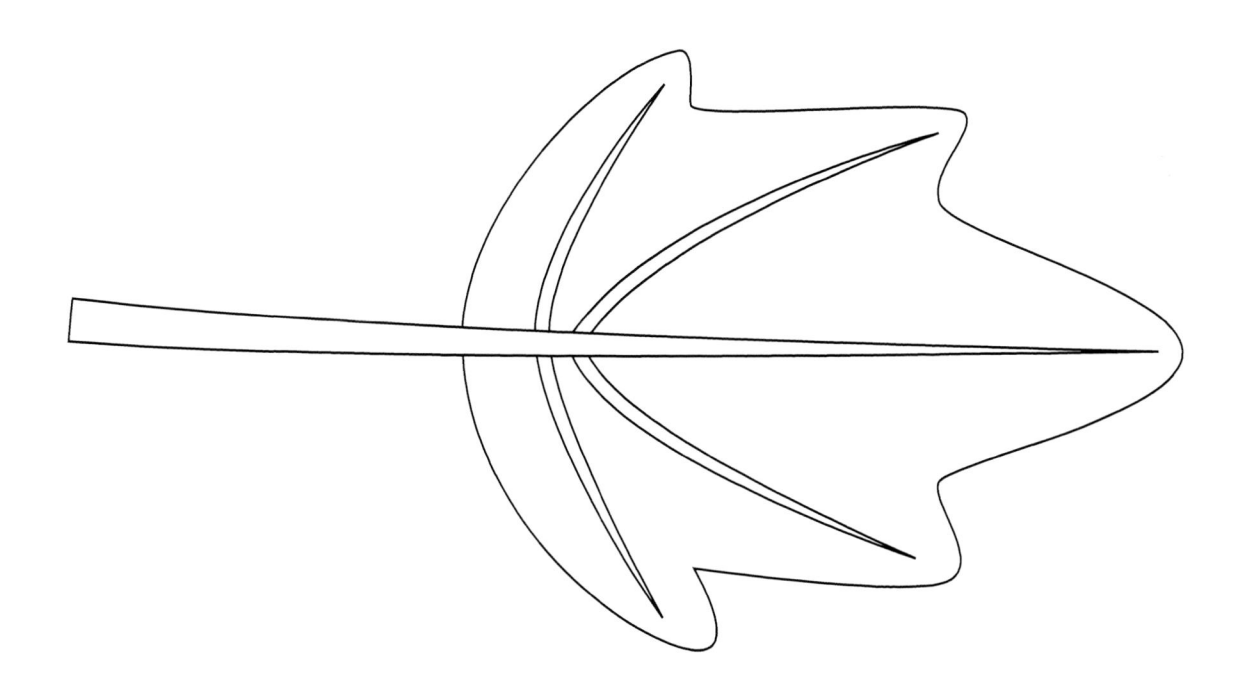

_____ inches

Measurement **Lesson** **4** Measure and Compare Lengths

Name _____

Measure the lengths and complete the sentence to compare.

1.

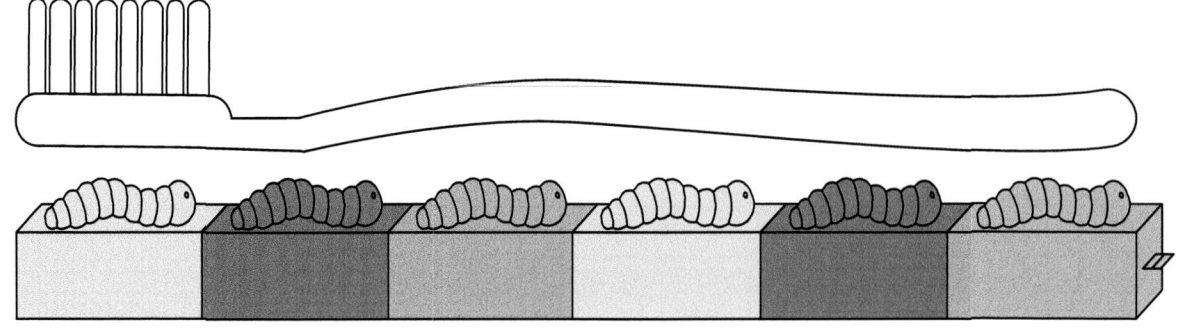

_____ inches

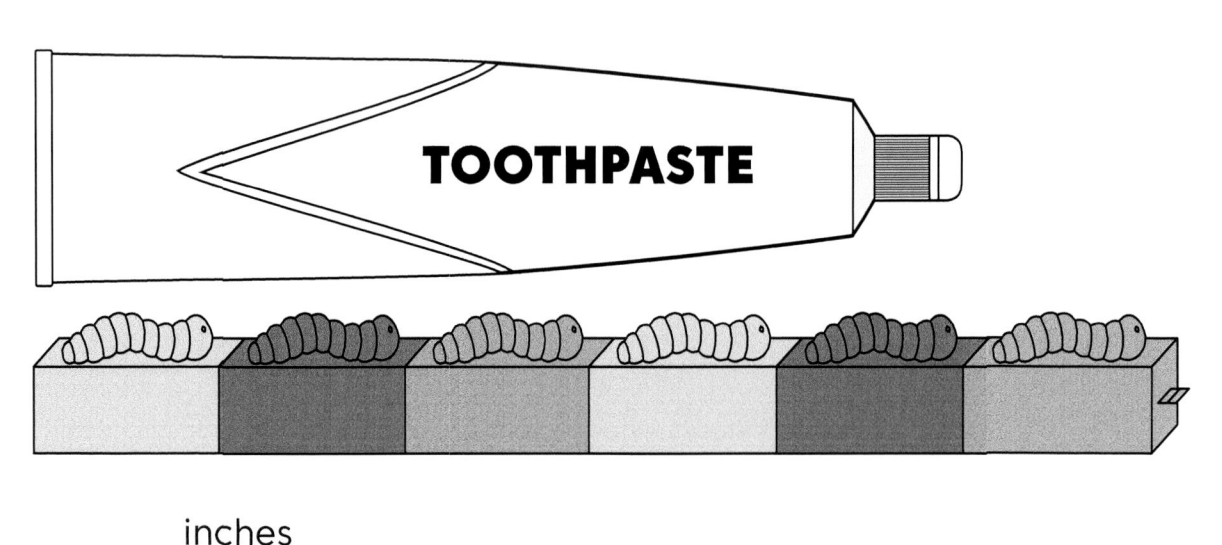

_____ inches

The toothbrush is _____ in. longer / shorter than the toothpaste.

Challenge! Isabella and Pablo want to compare the length of their pencils. Isabella uses the centimeter side of a ruler to measure her pencil. What should Pablo do?

Measurement Lesson 5 — Whole Numbers as Lengths on a Number Line

Name _____

1-cm Number Lines

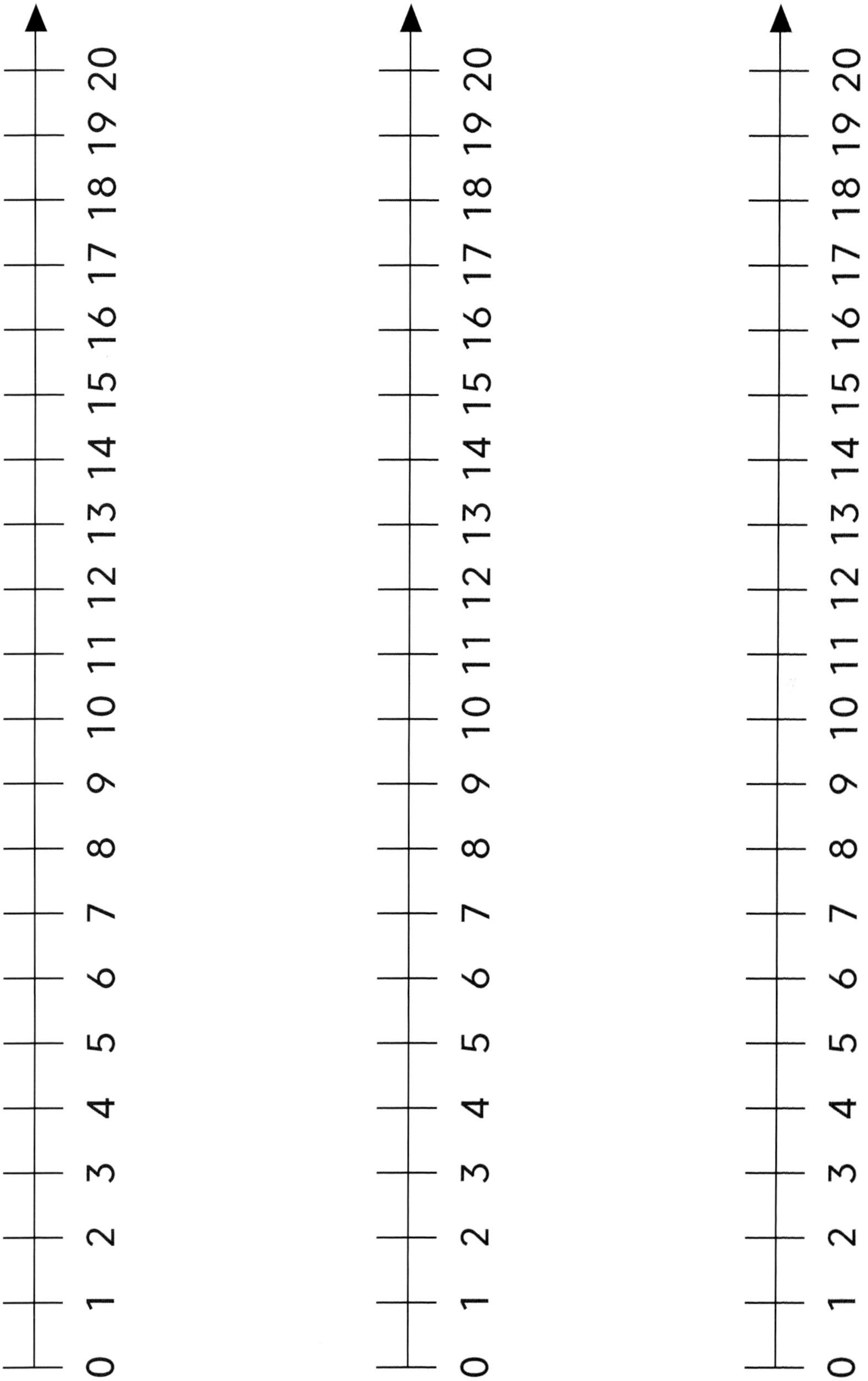

Measurement Lesson 5 — Whole Numbers as Lengths on a Number Line

Name _____

1-Inch Number Lines

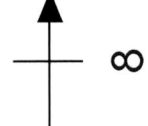

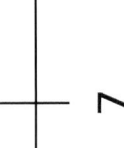

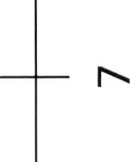

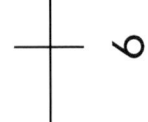

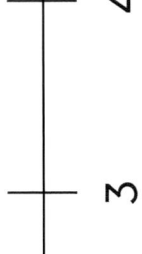

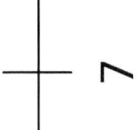

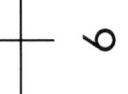

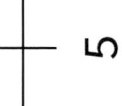

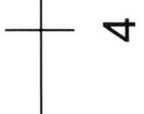

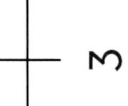

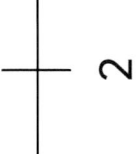

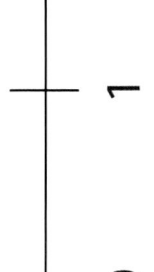

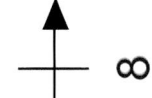

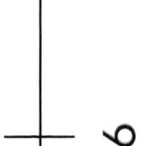

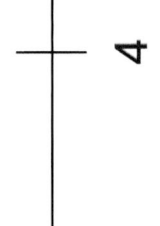

0 1 2 3 4 5 6 7 8

0 1 2 3 4 5 6 7 8

0 1 2 3 4 5 6 7 8

Measurement Lesson 5: Whole Numbers as Lengths on a Number Line

Name _____

Use Cuisenaire® Rods. Build the model. Write a number sentence for the lengths.

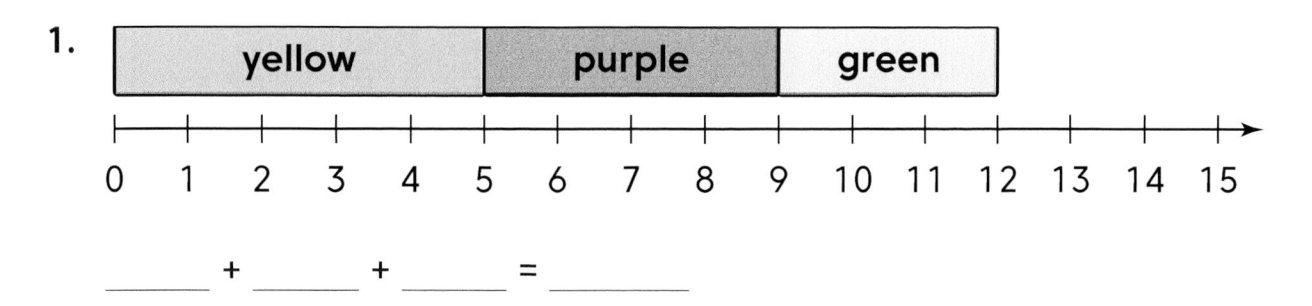

1. [yellow | purple | green]

   0 1 2 3 4 5 6 7 8 9 10 11 12 13 14 15

   ____ + ____ + ____ = ____

Use Cuisenaire® Rods. Build the length. Write a number sentence for the lengths.

2. 11 cm, two rods

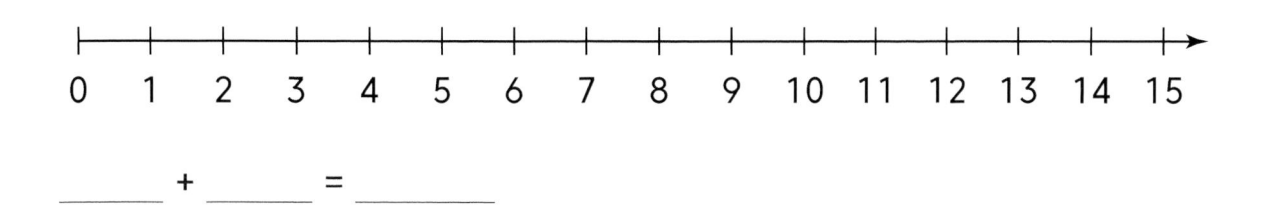

   0 1 2 3 4 5 6 7 8 9 10 11 12 13 14 15

   ____ + ____ = ____

Show the total length on the number line.

3. 3 + 1 + 6 + 5

   0 1 2 3 4 5 6 7 8 9 10 11 12 13 14 15

Challenge! Kyle's group uses 2 rods to form 8. Kyle uses 1 dark green rod and 1 red rod. Keisha uses 2 purple rods. Who is correct? Why?

Measurement **Lesson 6** Tell Time to 5 Minutes   Name _____

Use a NumberLine Clock™. Model the time shown.
Write the time.

1.

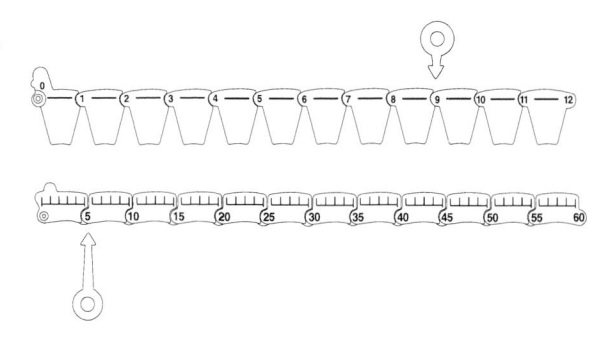

2.

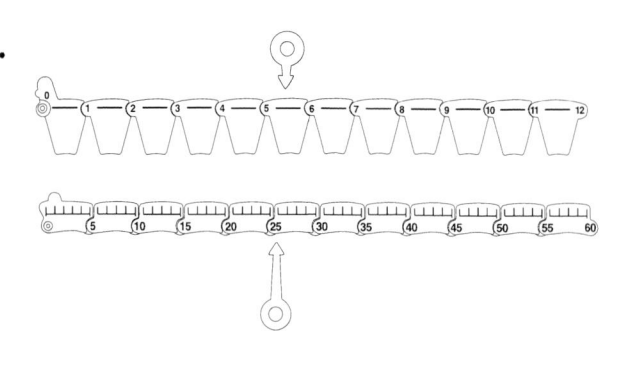

_____   _____

Use a NumberLine Clock™. Model the time shown.
Draw the hands on the clock.

3. 2:10

4. 8:50

Challenge! Kenji's mom says he can play video games until 3:10. He draws a clock to remind him of the time. The hour hand points between the red 3 and 4, closer to the 3. The minute hand points to the red 10. Does Kenji draw the clock correctly? Explain why or why not.

70   Hands-On Standards Math Intervention: Geometry, Measurement & Data   hand2mind.com

Measurement **Lesson 7** Tell Time to the Minute   Name _____

Use a NumberLine Clock™ to model the time. Write the time.

1. _____

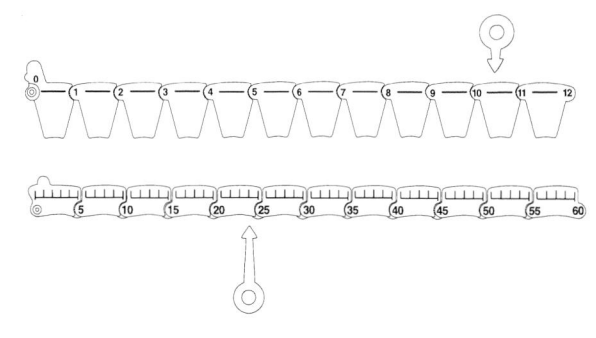

2. _____

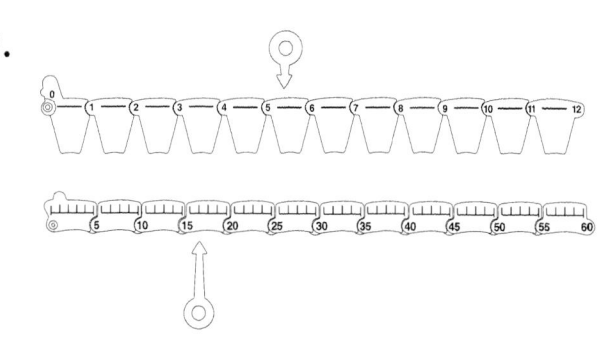

Use a NumberLine Clock™ to model the time. Draw the hands on the clock.

3. 11:04

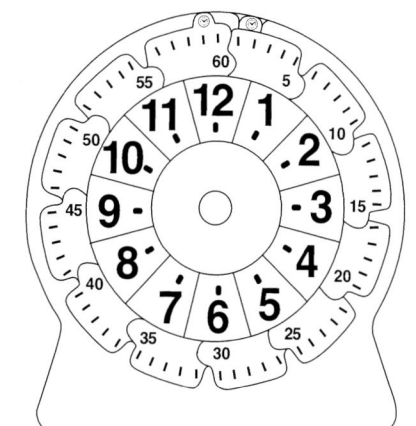

4. 2:52

Challenge! When Maya goes to her friend's house, the hour hand is halfway between the 3 and 4. Where will the hour hand be an hour later when she leaves?

Measurement Lesson 8: Solve Coin Problems

Name _____

Hundred Chart

1	2	3	4	5	6	7	8	9	10
11	12	13	14	15	16	17	18	19	20
21	22	23	24	25	26	27	28	29	30
31	32	33	34	35	36	37	38	39	40
41	42	43	44	45	46	47	48	49	50
51	52	53	54	55	56	57	58	59	60
61	62	63	64	65	66	67	68	69	70
71	72	73	74	75	76	77	78	79	80
81	82	83	84	85	86	87	88	89	90
91	92	93	94	95	96	97	98	99	100

Measurement
Lesson **8** Solve Coin Problems   Name _____

Circle the price. Color the amount paid. Find the change.

1. **price:** 53 cents; **paid:** 60 cents

1	2	3	4	5	6	7	8	9	10
11	12	13	14	15	16	17	18	19	20
21	22	23	24	25	26	27	28	29	30
31	32	33	34	35	36	37	38	39	40
41	42	43	44	45	46	47	48	49	50
51	52	53	54	55	56	57	58	59	60
61	62	63	64	65	66	67	68	69	70
71	72	73	74	75	76	77	78	79	80
81	82	83	84	85	86	87	88	89	90
91	92	93	94	95	96	97	98	99	100

_____

2. **price:** 77 cents; **paid:** 85 cents

1	2	3	4	5	6	7	8	9	10
11	12	13	14	15	16	17	18	19	20
21	22	23	24	25	26	27	28	29	30
31	32	33	34	35	36	37	38	39	40
41	42	43	44	45	46	47	48	49	50
51	52	53	54	55	56	57	58	59	60
61	62	63	64	65	66	67	68	69	70
71	72	73	74	75	76	77	78	79	80
81	82	83	84	85	86	87	88	89	90
91	92	93	94	95	96	97	98	99	100

_____

Find how much more is needed to buy each item.

3. **price:** 62 cents

   _____ is needed?

4. **price:** 47 cents

   _____ is needed?

5. Walid has 32 cents. He wants to buy a yogurt for 50 cents. How much more money does he need?

6. Daria buys a hair tie for 43 cents. She gives the clerk 50 cents. How much change should she get back?

Challenge! Kiara buys a drink for 36 cents. She gives the clerk 50 cents. The clerk gives her a dime and a nickel. Is that the correct change? Why?

Hands-On Standards Math Intervention: Geometry, Measurement & Data   hand2mind.com   73

Data
Lesson **1** Line Plots Using Inches   Name _____

Bookmark Line Plot Recording Sheet

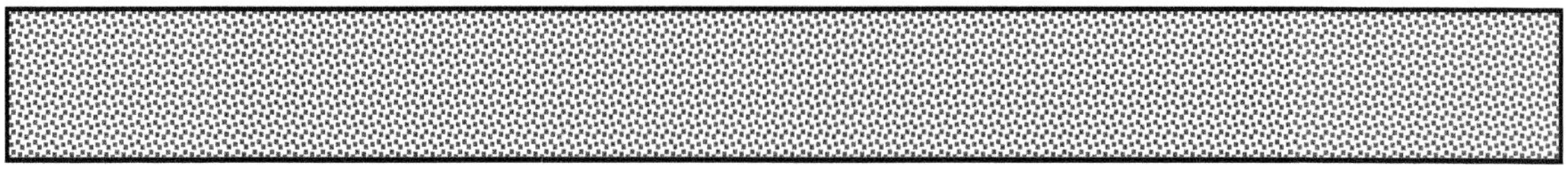

_____ inches

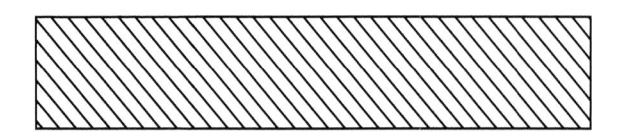

_____ inches

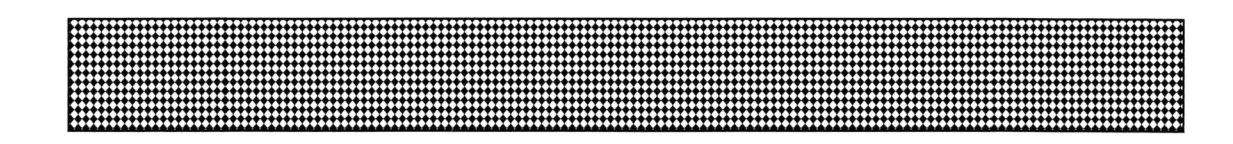

_____ inches

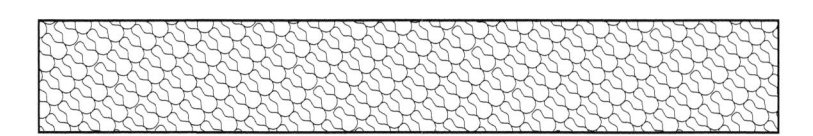

_____ inches

Number of Inches

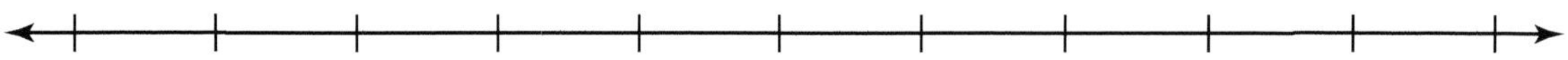

Data **Lesson** **1** Line Plots Using Inches   Name _____

Use the Inchworms™ to measure each pencil.

1. _____ inches

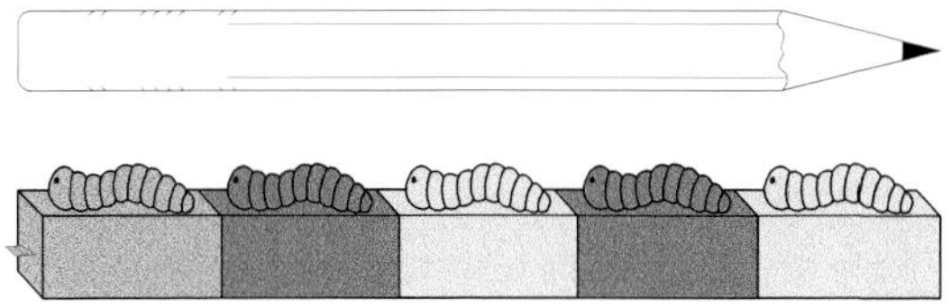

2. _____ inches

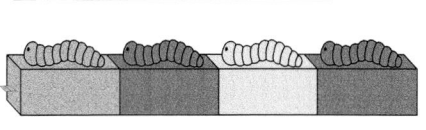

3. _____ inches

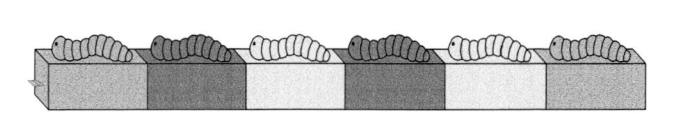

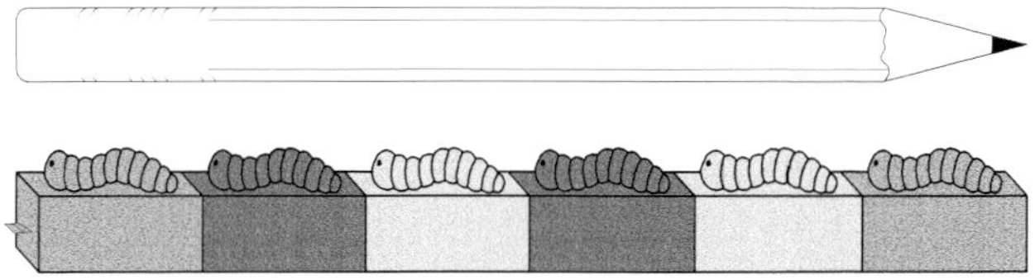

4. _____ inches

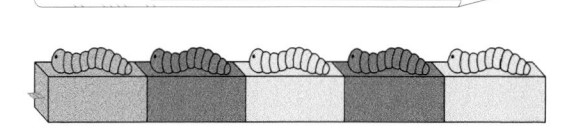

Use the lengths of the pencils to complete the line plot.

5. Number of Inches

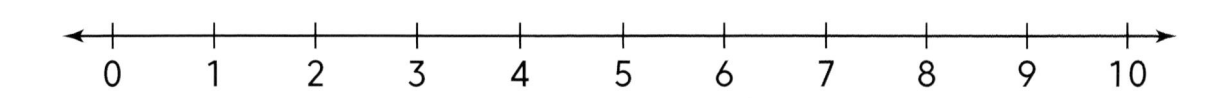

0 1 2 3 4 5 6 7 8 9 10

Challenge! Maya unwraps a new pack of 4 pencils. What would a line plot of the lengths of these pencils look like?

Data **Lesson** (2) Line Plots Using Centimeters

Name _____

Nail Line Plot Recording Sheet

_____ centimeters

_____ centimeters

_____ centimeters

_____ centimeters

Number of Centimeters

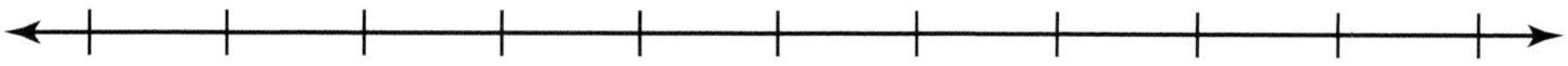

Data Lesson **2** Line Plots Using Centimeters

Name _____

Measure the length of each piece of yarn.

1. _____ centimeters

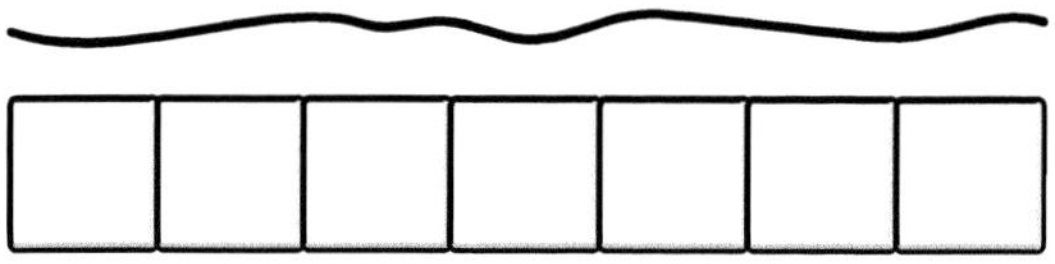

2. _____ centimeters

3. _____ centimeters

4. _____ centimeters

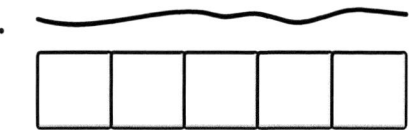

Make a line plot to show the lengths.

5. Number of centimeters

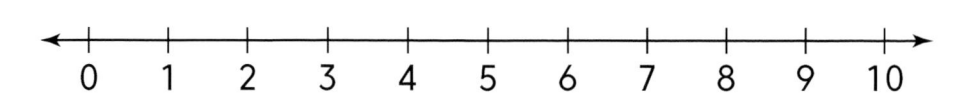

←——|——|——|——|——|——|——|——|——|——|——→
  0   1   2   3   4   5   6   7   8   9   10

Challenge! How could you use this line plot to help you make a craft?

Data **Lesson** ③ Solve Problems Using a Line Plot

Name _____

Line Plot Recording Sheet

Tally Chart	
1 in.	
2 in.	
3 in.	
4 in.	
5 in.	
6 in.	
7 in.	
8 in.	

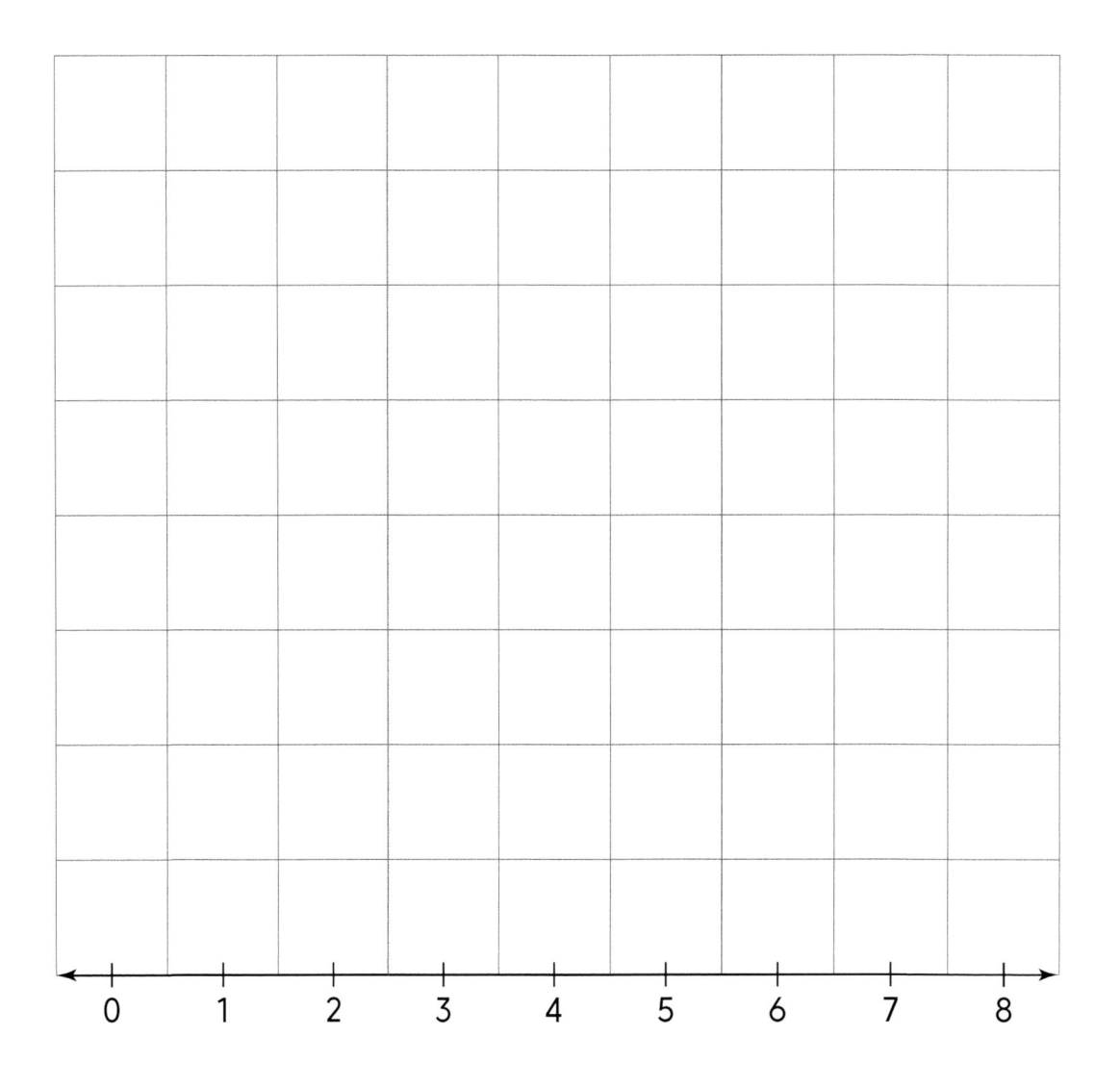

0   1   2   3   4   5   6   7   8

Data Lesson **3** Solve Problems Using a Line Plot

Name _____

Use the measurement of the food to complete the tally chart and line plot.

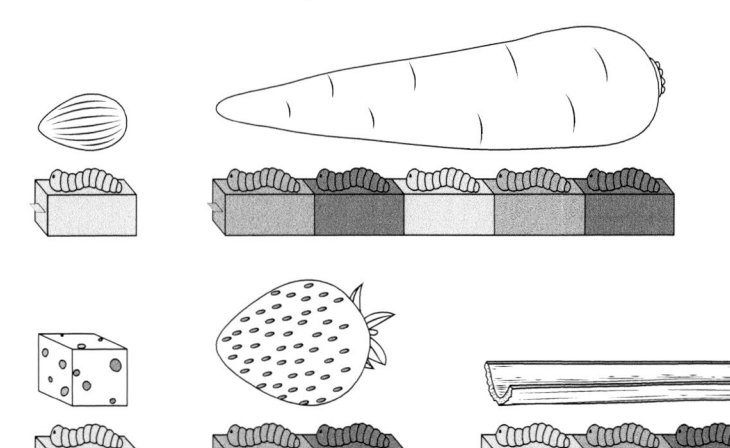

1. Complete the tally chart.

Tally Chart	
1 in.	
2 in.	
3 in.	
4 in.	
5 in.	

2. Use the tally chart to make a line plot.

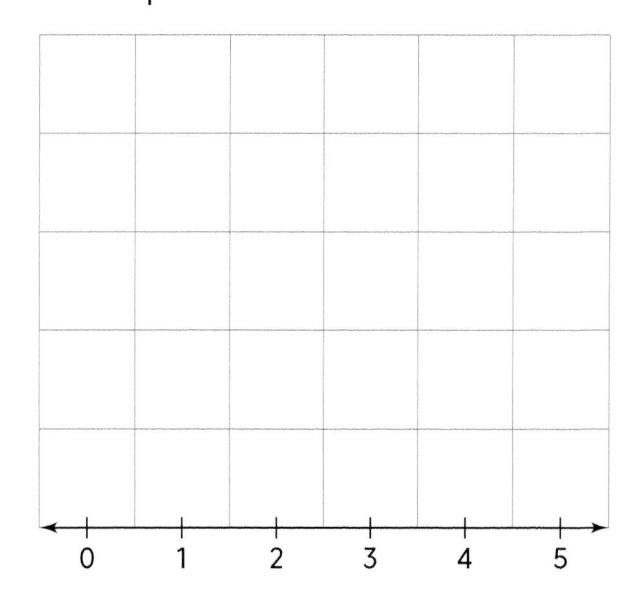

0 1 2 3 4 5

3. How many of the foods are shorter than 4 inches? _____

Challenge! Maya says that there can only be one X above each number because the X shows that there is at least one of those objects. Is Maya correct? Explain why or why not.

Hands-On Standards Math Intervention: Geometry, Measurement & Data   hand2mind.com   79

Data
Lesson 4 Picture Graphs

Name _____

Picture Graph Recording Sheet

Key: Each shape represents 1 solid.

Data **Lesson** **4** Picture Graphs Name _____

Use the solids to complete the picture graphs.

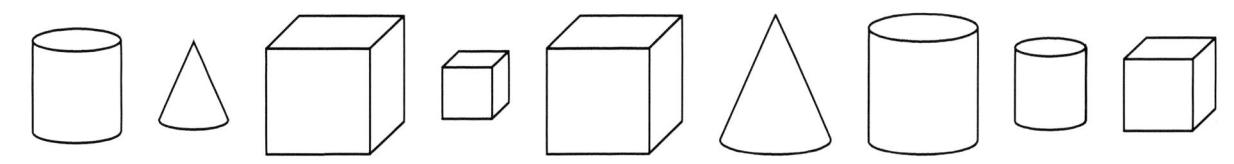

1.

Solid Shapes				
Cylinders				
Cubes				
Cones				

Key: Each shape represents 1 solid.

Challenge! Suppose there were two more cubes. How would the picture graph change? How would it stay the same?

Data
Lesson 5 Bar Graphs

Name _____

Bar Graph Recording Sheet

_____   _____   _____   _____

Data Lesson **5** Bar Graphs   Name _____

Use the Coin Tiles to count the number of coins.

1.

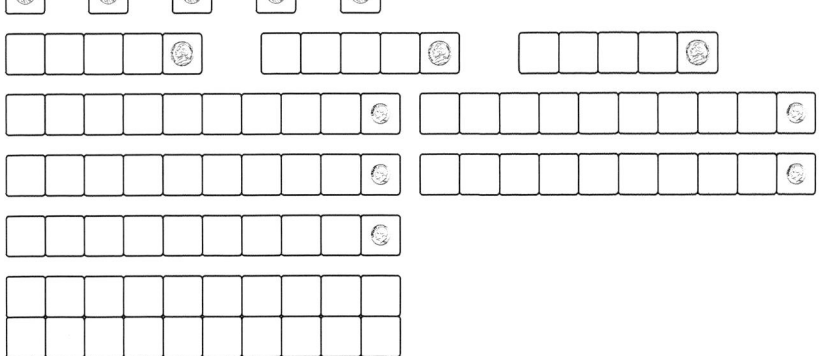

_____ pennies   _____ nickels   _____ dimes   _____ quarters

Use the numbers of coins to complete the bar graph.

2.

Use the bar graph to answer the question.

3. Which coin is there the least of? _____

4. How many pennies and nickels are there together? _____

Challenge! How would the graph change if the 5 pennies are exchanged for 1 nickel?

Hands-On Standards Math Intervention: Geometry, Measurement & Data   hand2mind.com   83

Data
Lesson 6 Solve Problems Using Graphs

Name _____

Insect Recording Sheet

_____   _____   _____   _____

Data Lesson **6** Solve Problems Using Graphs

Name _____

The bar graph shows the number of students and how they get to school.

How Students Get to School

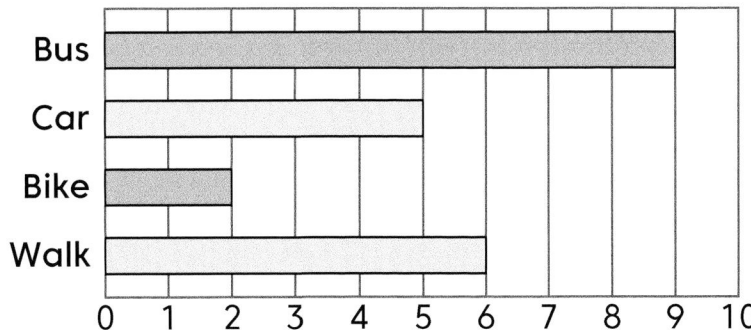

Bus	9
Car	5
Bike	2
Walk	6

(scale: 0 1 2 3 4 5 6 7 8 9 10)

1. What way do the most students get to school? _____

2. How many students ride the bus or a car? _____ students

The picture graph shows the favorite sport of each student in Mr. Ramirez's class.

Favorite Sport			
⚽			
⚽			⚾
⚽		🏀	⚾
⚽	🏈	🏀	⚾
⚽	🏈	🏀	⚾
Soccer	Football	Basketball	Softball

Key: Each ball represents 1 student.

3. How many more students chose soccer than basketball as their favorite sport?

Challenge! On a rainy day, 3 students in Ms. Barclay's class who usually walk rode in cars. How would this change the How Students Get to School graph?

Multi-Lesson Blackline Master 1: 1-Inch Grid Paper

Name _____

Multi-Lesson **Blackline Master** (2) Telling Time Recording Sheet

Name _____

1. _____ : _____

2. _____ : _____

3. _____ : _____

4. _____ : _____

Hands-On Standards Math Intervention: Geometry, Measurement & Data

hand2mind.com **87**

Assessment **Grade** (2) Geometry Assessment Name _____

1. What is the name of the shape?

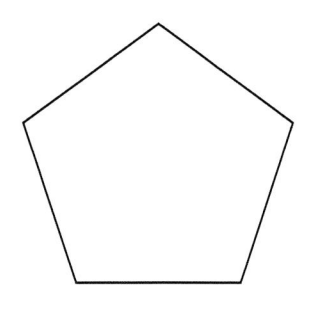

   _____

2. Ali says both of these shapes are triangles. Do you agree? Explain why or why not.

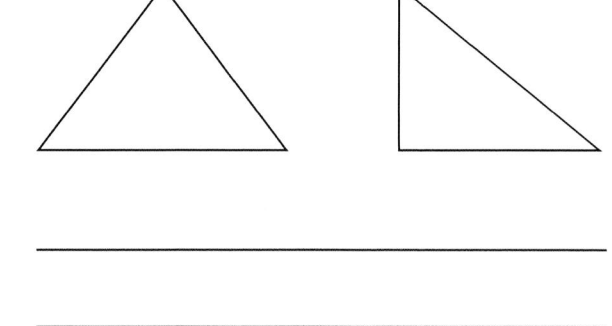

   _____
   _____
   _____
   _____

3. Draw lines to make 4 equal parts.

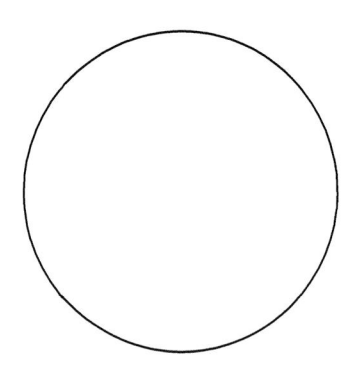

4. Minh uses Color Tiles to build a rectangle with 3 rows and 6 columns. How many tiles does she use?

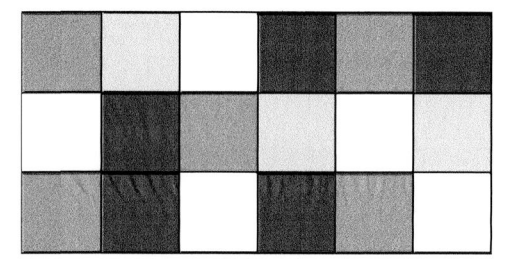

   _____

Assessment Grade 2 — Geometry Assessment   Name _____

5. Fill in the blanks to complete the sentence about the circle.

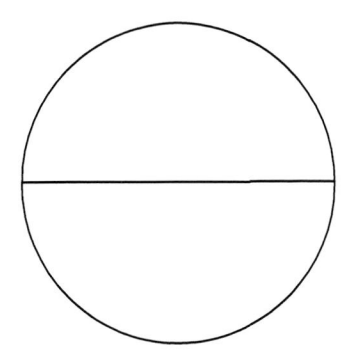

Each part of the circle is a _____.

The whole circle is made up of _____ halves.

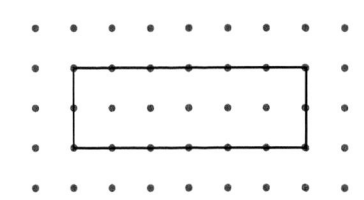

6. Lupe uses Color Tiles to build a rectangle with 2 rows and 6 columns. Draw the shape. Write a number sentence to find how many tiles were used.

_____

7. Draw a line to make 2 equal parts.

_____

8. What is the name for a shape with 6 sides?

_____

Hands-On Standards Math Intervention: Geometry, Measurement & Data   hand2mind.com   89

Assessment **Grade 2** Geometry Assessment   Name _____

9. The rectangle is partitioned into 3 equal parts. What is each part of the whole called?

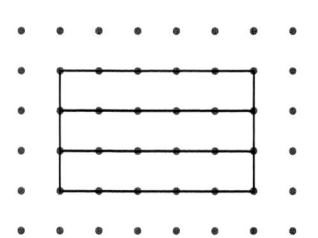

_____

10. Jeremy uses Color Tiles to build a rectangle with 5 rows and 4 columns. Write a number sentence to find how many tiles were used.

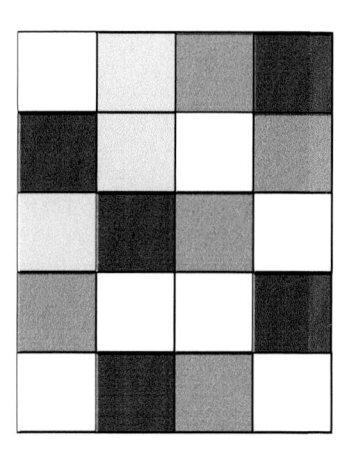

_____

Assessment Grade **2** — Measurement Assessment   Name _____

1. Estimate the length of the screwdriver in inches and centimeters.

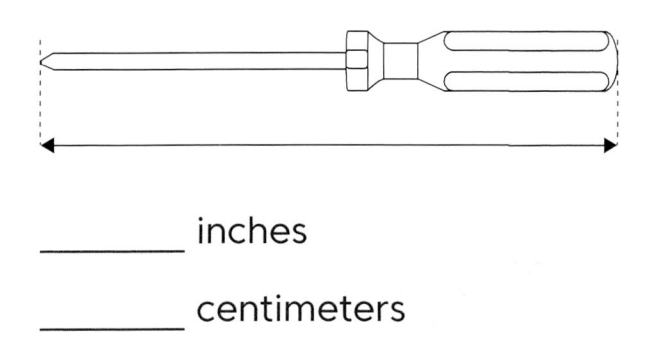

   _____ inches

   _____ centimeters

2. Measure the crayon in inches and centimeters.

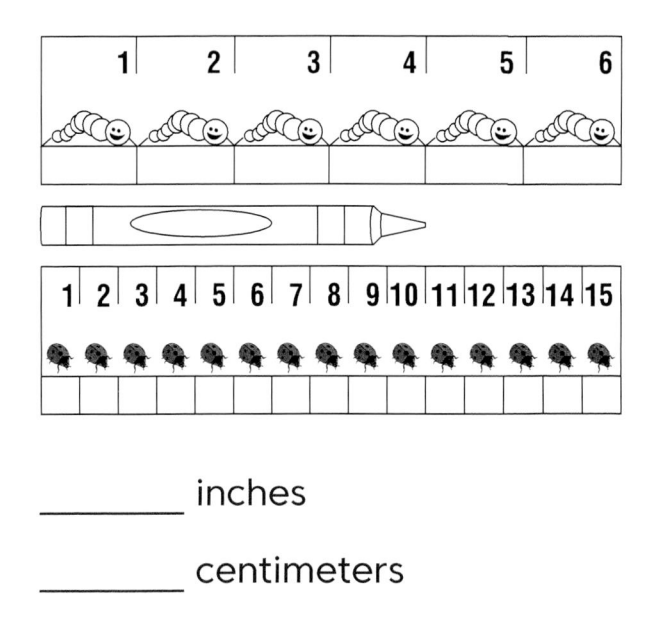

   _____ inches

   _____ centimeters

3. Measure the lengths and complete the sentence to compare.

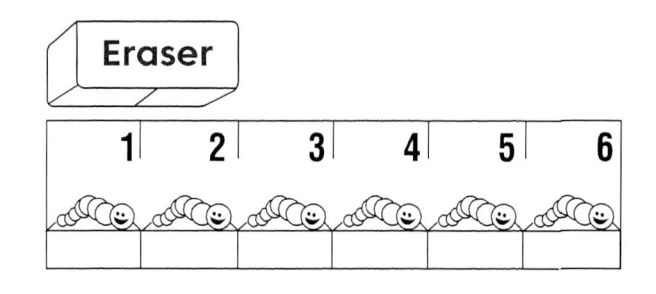

   Eraser

   _____ inches

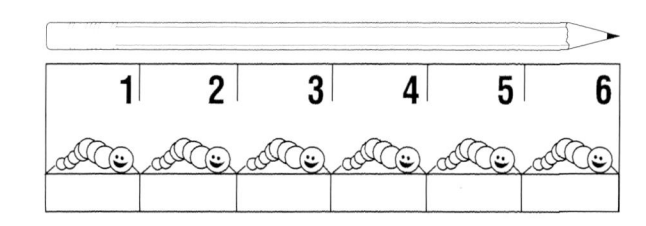

   _____ inches

   The pencil is _____ inches longer than the eraser.

4. Write a number sentence for the lengths shown on the number line.

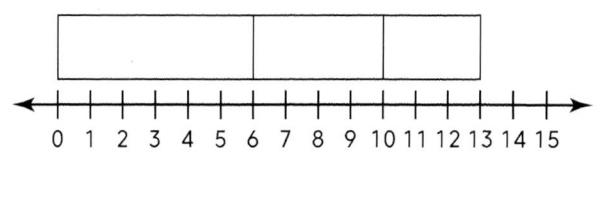

   _____ + _____ + _____ = _____

Hands-On Standards Math Intervention: Geometry, Measurement & Data   hand2mind.com

Assessment **Grade 2** — Measurement Assessment   Name _____

5. Show the total length on the number line.

3 + 5 + 1 + 2

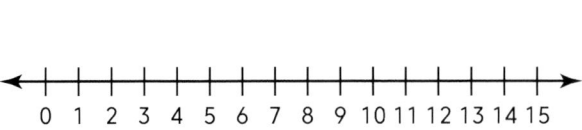

```
←|—|—|—|—|—|—|—|—|—|—|—|—|—|—|—|→
  0 1 2 3 4 5 6 7 8 9 10 11 12 13 14 15
```

6. Model the time 7:12. Draw the hands on the clock.

7. How many more cents are needed to buy an item that costs 38 cents?

_____

8. Sari measures a door. Should she use inches or feet to measure? What tool should she use? Explain

_____
_____
_____

9. Write the time shown on the clock.

_____

10. Andrea buys a gumball for 35 cents. She gives the clerk 50 cents. How much change should she get back?

_____

92   Hands-On Standards Math Intervention: Geometry, Measurement & Data   hand2mind.com

Assessment **Grade** **2** Data Assessment

Name _____

1. Use the Inchworms™ to measure each stick. Then make a line plot to show the lengths.

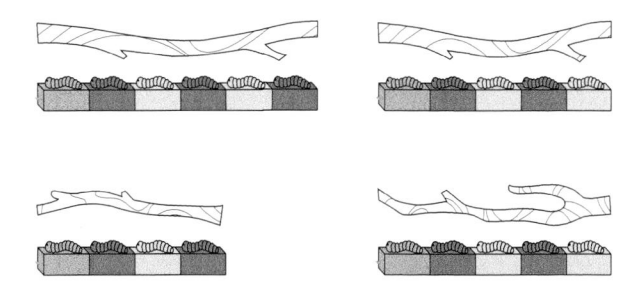

Number of Inches

0 1 2 3 4 5 6 7 8 9 10

2. Complete the picture graph to show the fruit.

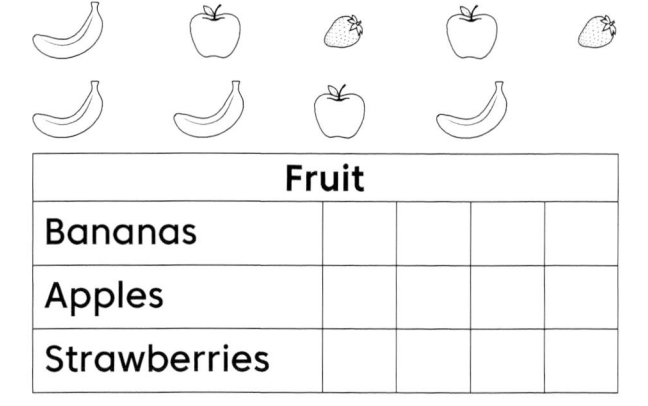

Fruit			
Bananas			
Apples			
Strawberries			

3. Measure the length of each caterpillar. Then make a line plot to show the lengths.

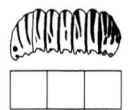

_____ centimeters

_____ centimeters

_____ centimeters

_____ centimeters

Number of Centimeters

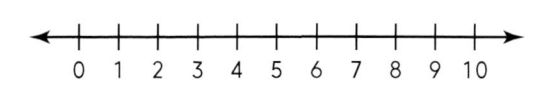

0 1 2 3 4 5 6 7 8 9 10

Hands-On Standards Math Intervention: Geometry, Measurement & Data

Assessment Grade 2 Data Assessment

Name _____

4. The bar graph shows the number of students and their favorite food.

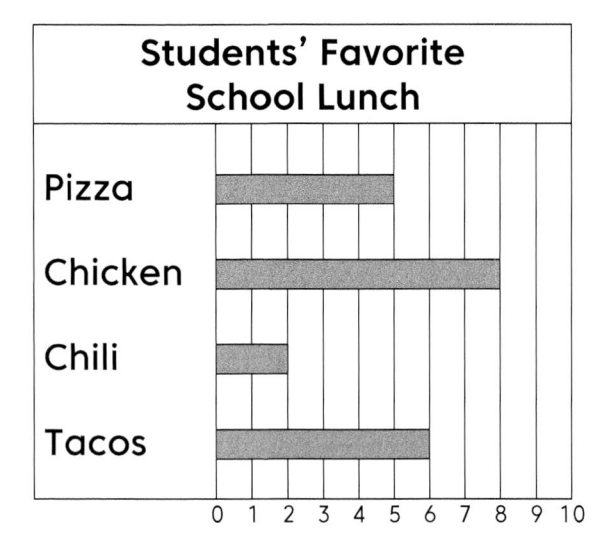

Students' Favorite School Lunch

Pizza, Chicken, Chili, Tacos
0 1 2 3 4 5 6 7 8 9 10

How many students prefer chicken and tacos?

_____

5. Use the leaves to complete the tally chart. Then make a line plot to show the lengths.

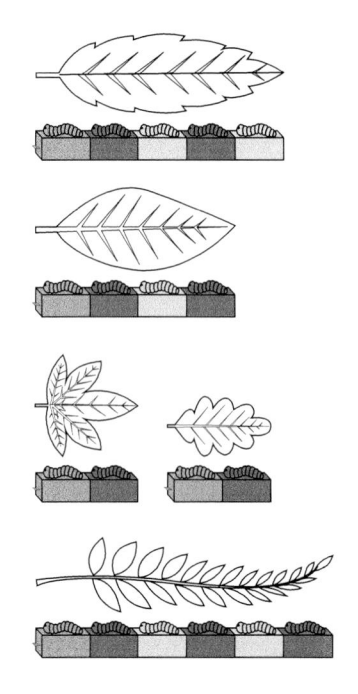

Tally Chart	
1 in.	
2 in.	
3 in.	
4 in.	
5 in.	
6 in.	

Length of Leaves in Inches

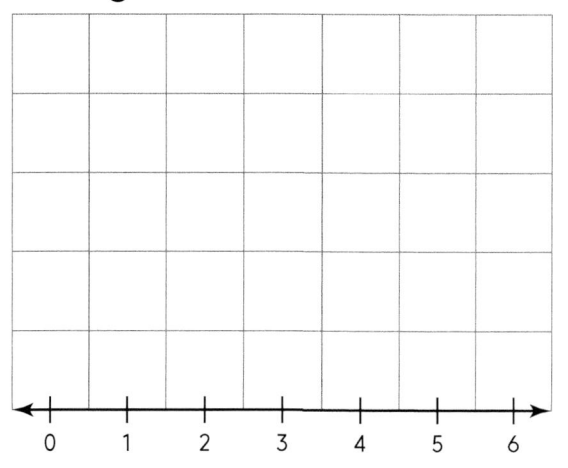

0 1 2 3 4 5 6

Assessment **Grade 2** — Data Assessment

Name _____

6. The tally chart shows the lengths of some baby lizards. Use the tally chart to make a line plot.

Tally Chart	
1 in.	
2 in.	
3 in.	||
4 in.	|||
5 in.	||
6 in.	|||

Length of Baby Lizards in Inches

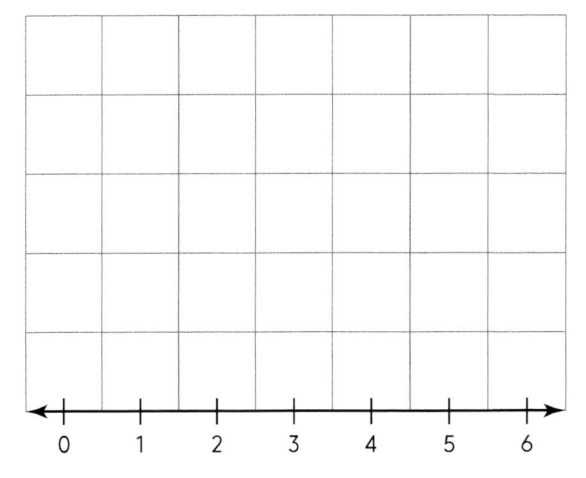

0 1 2 3 4 5 6

7. Use the toys to complete the bar graph. Then use the bar graph to answer a question.

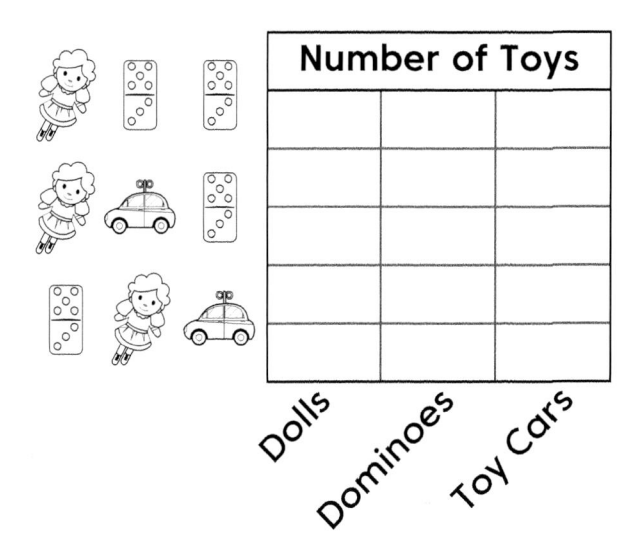

Number of Toys

Dolls   Dominoes   Toy Cars

Which toy is there the most of?

_____

8. Mari has four crayons. The crayons measure 7 cm, 8 cm, 6 cm, and 7 cm. Make a line plot to show the lengths.

Number of Centimeters

0 1 2 3 4 5 6 7 8 9 10

Assessment Grade 2 — Data Assessment

Name _____

9. Use the shapes to complete the picture graph.

○ △ ○ □
△ ○ □ △

Shape				
Circles				
Squares				
Triangles				

10. Rosario goes to the park and counts how many people she sees doing different activities. The graph shows her data.

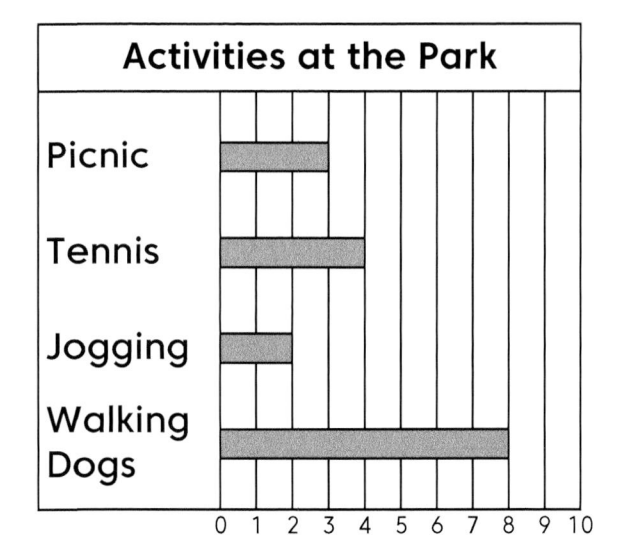

Activities at the Park

Picnic — 3
Tennis — 4
Jogging — 2
Walking Dogs — 8

0 1 2 3 4 5 6 7 8 9 10

How many more people are walking dogs than playing tennis?

_____ people

Grade 2 Assessment Student Progress Report

Name _____

Use this sheet to record assessment results for each student.

Unit 1: Geometry

Item No.	Lesson No.	Concept	Meets? Y/N __/__/__	__/__/__	__/__/__
1.	1	Classify a shape			
2.	1	Classify a shape			
3.	6	Partition a circle into equal fourths			
4.	3	Find total squares			
5.	6	Identify halves			
6.	4	Write a repeated addition sentence			
7.	5	Partition a rectangle			
8.	2	Attributes of shapes			
9.	5	Name partitioned parts			
10.	4	Write a repeated addition sentence			

Unit 2: Measurement

Item No.	Lesson No.	Concept	Meets? Y/N __/__/__	__/__/__	__/__/__
1.	1	Estimate length			
2.	2	Measure an object			
3.	4	Compare lengths			
4.	5	Represent sums as lengths			
5.	5	Represent sums as lengths			
6.	7	Time to the minute			
7.	8	Money			
8.	3	Select tool to measure			
9.	6	Time to 5 minutes			
10.	8	Money			

Hands-On Standards Math Intervention: Geometry, Measurement & Data

hand2mind.com

Grade 2 Assessment Student Progress Report

Name _____

Use this sheet to record assessment results for each student.

Unit 3: Data					
Item No.	Lesson No.	Concept	Meets? Y/N		
			__/__/__	__/__/__	__/__/__
1.	1	Create a line plot			
2.	4	Create a picture graph			
3.	2	Create a line plot			
4.	6	Solve problems using data			
5.	3	Use a tally chart			
6.	3	Use a tally chart			
7.	5	Create a bar graph			
8.	2	Create a line plot			
9.	4	Create a picture graph			
10.	6	Solve problems using data			

Grade 2 Answer Key

Geometry
Lesson 1: Identify Shapes
Formative Assessment

quadrilateral/trapezoid

1. quadrilateral
2. triangle
3. quadrilateral
4. pentagon

Challenge

Yes, they can both be correct because a rectangle is also a quadrilateral.

Lesson 2: Recognize and Draw Shapes
Formative Assessment

triangular pyramid

1. hexagon
2. rectangular prism
3. triangle
4. pentagon

Challenge

Possible answers: cereal boxes, juice boxes, laptops, pizza boxes, notebooks, books, or any other rectangular-prism-shaped everyday object

Lesson 3: Partition Rectangles
Formative Assessment

12 tiles

1. 14
2. 6
3. 10
4. 4
5. 12
6. 15

Challenge

Possible answers: 1 row, 18 columns; 18 rows, 1 column; 9 rows, 2 columns; 2 rows, 9 columns; 3 rows, 6 columns; or 6 rows, 3 columns

Lesson 4: Solve Problems by Partitioning Rectangles
Formative Assessment

27 tiles

1. 10 + 10 + 10 = 30
2. 5 + 5 + 5 + 5 + 5 + 5 = 30
3. 8 + 8 = 16
4. 3 + 3 + 3 + 3 + 3 + 3 + 3 + 3 + 3 = 27
5. 2 + 2 + 2 + 2 + 2 = 10
6. 7 + 7 + 7 + 7 = 28

Challenge

Possible answers: 1 row, 24 columns; 24 rows, 1 column; 2 rows, 12 columns; 12 rows, 2 columns; 3 rows, 8 columns; 8 rows, 3 columns; 4 rows, 6 columns; or 6 rows, 4 columns

With 4 fewer placemats: 1 row, 20 columns; 20 rows, 1 column; 2 rows, 10 columns; 10 rows, 2 columns; 4 rows, 5 columns; 5 rows, 4 columns

Lesson 5: Partition Rectangles into Fair Shares
Formative Assessment

Each part is called a half.

1. halves
2. thirds
3. half
4. third

Challenge

Check students' drawings. One possible answer could be a rectangle that is 3 grid squares by 2 grid squares, showing one partition into rows and the other into columns.

Lesson 6: Partition Circles
Formative Assessment

Each part is called a third.

1. The circle divided into halves.
2. The circle divided into thirds.
3. Check students' drawings.
4. Check students' drawings.

Challenge

Each part is 1 half. Together, there are 2 halves.

Measurement
Lesson 1: Estimate Lengths
Formative Assessment

Possible answer: I estimated that it was 2 inches. It measured 2 inches.

1. Check students' estimates; 4 inches; 10 centimeters
2. Check students' estimates; 3 inches; 8 centimeters

Grade 2 Answer Key

Challenge

No, 3 centimeters would be too small of a pencil to hold.

Lesson 2: Different Size Units

Formative Assessment

Possible answer: The two measurements are different because inches are larger than centimeters.

1. 5 inches, 13 centimeters

Challenge

Abbey measured in inches, and Jacob measured in centimeters. I know because centimeters are smaller, so it takes more of them to make the same length.

Lesson 3: Select and Use Measurement Tools

Formative Assessment

Possible answer: Asa could use an Inchworms™ and Centibugs™ Ruler.

1. InchWorms™; 1 inch
2. Inchworms™/Inchworms™ Ruler; 5 in.

Challenge

Sarika could use Inchworms™ or rulers to measure the macaroni and a string.

Lesson 4: Measure and Compare Lengths

Formative Assessment

Answers may vary

1. 6 inches; 5 inches; The toothbrush is 1 in. longer than the toothpaste.

Challenge

Pablo needs to use the centimeter side of the ruler as well so they can compare their measurements.

Lesson 5: Whole Numbers as Lengths on a Number Line

Formative Assessment

15 cm

1. 5 + 4 + 3 = 12
2. dark green rod and yellow rod, 6 + 5 = 11
3. green rod, white rod, dark green rod, yellow rod

Challenge

They are both correct. 6 + 2 = 8 and 4 + 4 = 8.

Lesson 6: Tell Time to 5 Minutes

Formative Assessment

3:45

1. 9:05
2. 5:25
3. 2:10
4. 8:50

Challenge

No, he did not draw the time correctly. The minute hand should point to the red 2, which is the same as the blue 10, to show it is 10 minutes after 3.

Lesson 7: Tell Time to the Minute

Formative Assessment

4:47 (on digital)

1. 10:24
2. 5:17
3.

Grade 2 Answer Key

4.

Challenge
Since it is 1 hour later, the hour hand will be halfway between the 4 and 5.

Lesson 8: Solve Coin Problems
Formative Assessment
10 cents

1. 7 cents
2. 8 cents
3. 12 cents
4. 7 cents
5. 18 cents
6. 7 cents

Challenge
She did not get the right change. The clerk gave her extra. She should have gotten 14 cents back, but the clerk gave her 15 cents.

Data

Lesson 1: Line Plots Using Inches
Formative Assessment

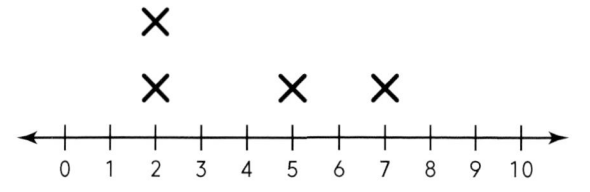

Number of Inches

1. 5
2. 4
3. 6
4. 5

5.

Number of Inches

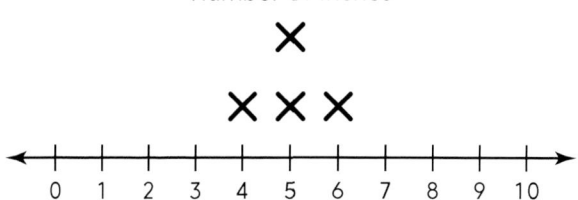

Challenge
It would have 4 Xs in the same column because they are all the same size.

Lesson 2: Line Plots Using Centimeters
Formative Assessment

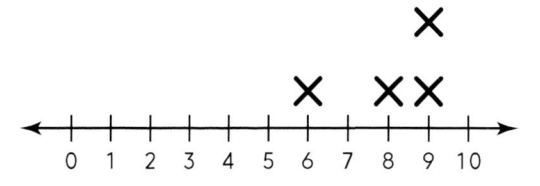

Number of Centimeters

1. 7
2. 3
3. 3
4. 5

5.

Number of Centimeters

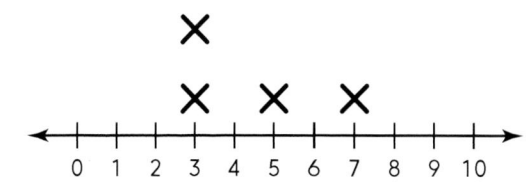

Challenge
You could use it to keep track of the number of each length of yarn that you cut.

Lesson 3: Solve Problems Using a Line Plot
Formative Assessment
4 lengths of fabric

1.

Tally Chart	
1 in.	II
2 in.	I
3 in.	I
4 in.	
5 in.	I

Grade 2 Answer Key

2.

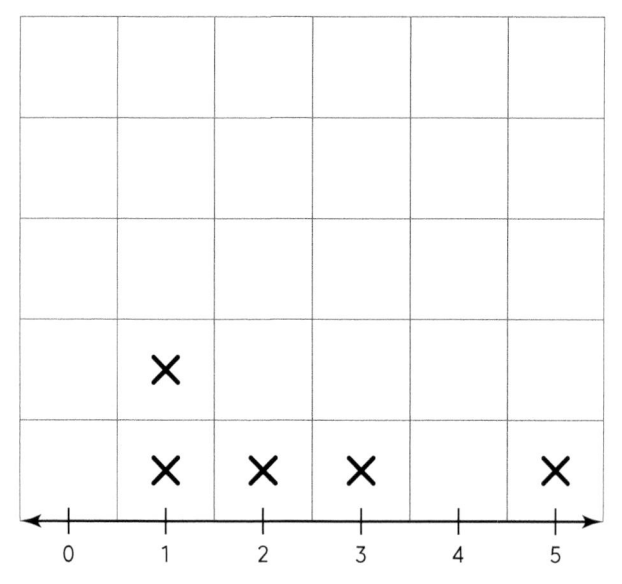

3. 4

Challenge
Maya is incorrect. Just like on a tally chart, you put a tally each time you count an object, and you should put a mark each time you have that measurement.

Lesson 4: Picture Graphs
Formative Assessment

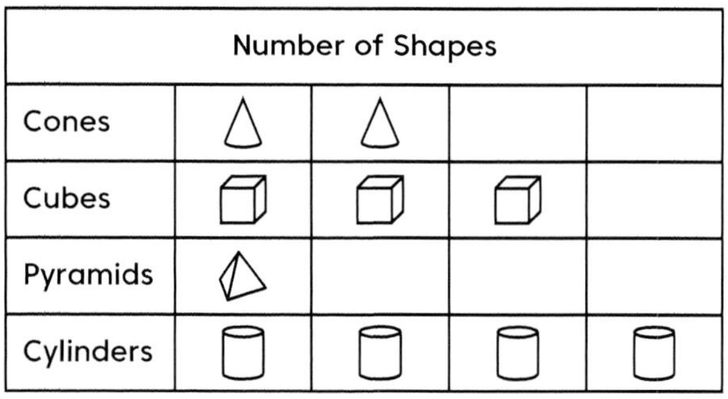

Number of Shapes				
Cones	△	△		
Cubes	⬜	⬜	⬜	
Pyramids	△			
Cylinders	⌭	⌭	⌭	⌭

Key: Each shape represents 1 solid.

1.

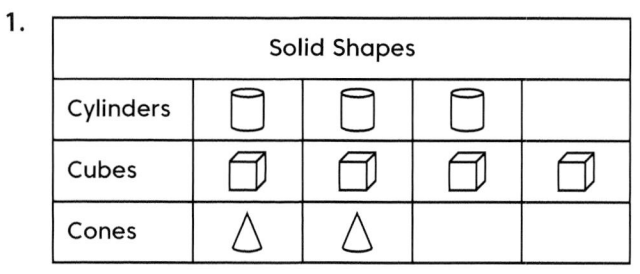

Solid Shapes				
Cylinders	⌭	⌭	⌭	
Cubes	⬜	⬜	⬜	⬜
Cones	△	△		

Key: Each shape represents 1 solid.

Challenge
Possible answer: In the picture graph, the row for cubes would have 2 more pictures. I would need to add 2 more columns. The pictorial representation for cylinders and cones will remain the same.

Lesson 5: Bar Graphs
Formative Assessment

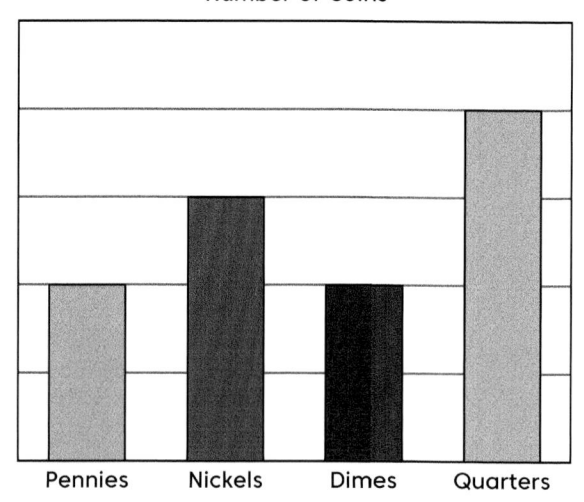

1. 5 pennies, 3 nickels, 5 dimes, 1 quarter

2.

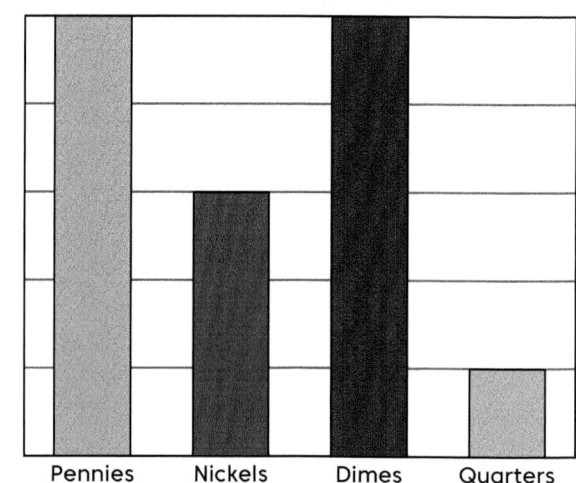

3. quarters

4. 8

Challenge
The bar for pennies would be at 0, and the height of the bar for nickels would go up by 1 to have a height of 4.

Lesson 6: Solve Problems Using Graphs
Formative Assessment
4

1. bus
2. 14 students
3. 2 students

Challenge
There would now be 3 students who walk and 8 students who ride in cars.

102   Hands-On Standards Math Intervention: Geometry, Measurement & Data   hand2mind.com

Grade 2 Answer Key

Assessment Answer Key

Geometry

1. pentagon
2. I agree because they both have 3 sides and 3 vertices.
3. The circle should be divided into 4 equal parts.
4. 18 tiles
5. half, 2
6. 6 + 6 = 12 tiles;

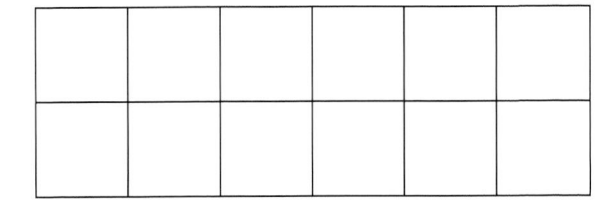

7. Students may divide the rectangle in half vertically or horizontally.
8. hexagon
9. one-third
10. 4 + 4 + 4 + 4 + 4 = 20

Measurement

1. 3, 8
2. 4, 10
3. 2, 6, 4
4. 6 + 4 + 3 = 13
5. Students' number lines should show 3 + 5 + 1 + 2 = 11
6. Short hand a little to the left of 7, long hand to the tick mark for 12 minutes
7. 3 cents
8. Possible answer: Feet because the door is large. She could use a ruler to measure the door.
9. 4:15
10. 15 cents

Data

1. Number of Inches

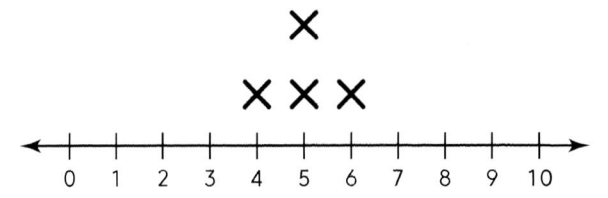

2.

Fruit				
Bananas	🍌	🍌	🍌	🍌
Apples	🍎	🍎	🍎	
Strawberries	🍓	🍓		

3. 3, 7, 3, 5

   Number of Centimeters

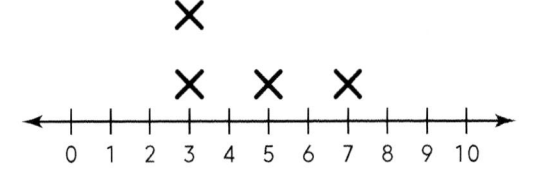

4. 14.

5.

Tally Chart	
1 in.	
2 in.	| |
3 in.	
4 in.	|
5 in.	|
6 in.	|

Length of Leaves in Inches

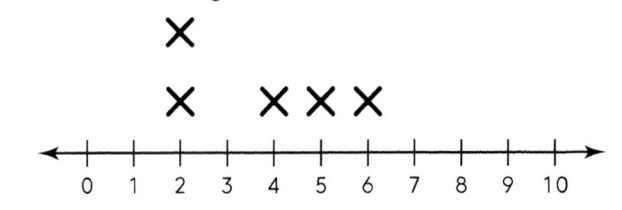

6. Length of Baby Lizards in Inches

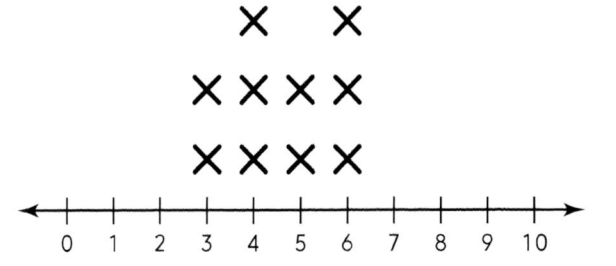

Grade ② Answer Key

7. Dominoes

Number of Toys

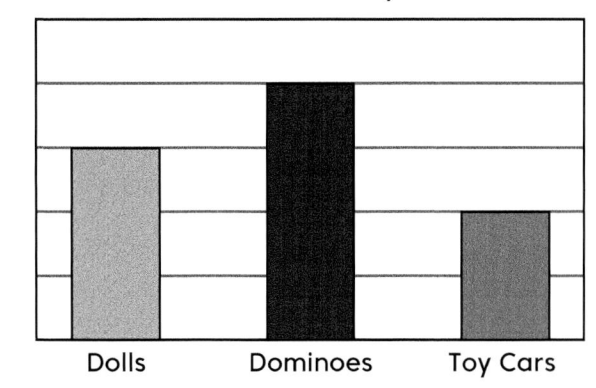

(bar graph: Dolls, Dominoes, Toy Cars)

8. Number of Centimeters

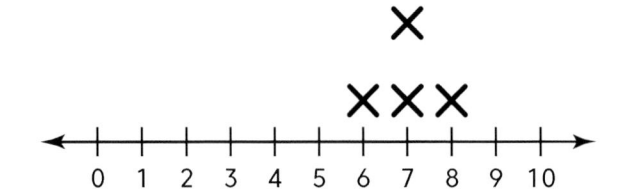

(line plot with X at 7, and X X X at 6, 7, 8 on number line 0–10)

9.

Shape				
Circles	○	○	○	
Squares	□	□		
Triangles	△	△	△	

10. 4

Glossary

Coin Tiles

These easy-to-handle tiles help students visualize the relationships between the values of pennies, nickels, dimes, and quarters. Each tile is a hands-on area model of the coin it represents. The tiles help students see meaning in the relationships between coins and understand concepts such as coin recognition, coin values, coin equivalence, and making change. Students can practice manipulating parts of a dollar by using the tiles with a hundred board.

Color Tiles

These 1-inch square tiles come in four different colors: red, blue, yellow, and green. They can be used to explore many mathematical concepts, including geometry, patterns, and number sense.

Cuisenaire® Rods

Cuisenaire® Rods include rods of 10 different colors, each corresponding to a specific length. White rods, the shortest, are 1 cm long. Orange rods, the longest, are 10 cm long. Rods allow students to explore all fundamental math concepts, including addition and patterning, multiplication, division, fractions and decimals, and data analysis.

Rainbow Fraction® Circles

A set of Rainbow Fraction® Circles includes 9 color coded, plastic circles representing a whole, halves, thirds, fourths, fifths, sixths, eighths, tenths, and twelfths. The circles enable children to explore fractions, fractional equivalences, the fractional components of circle graphs, and more.

Geoboard

The double-sided Geoboard is 7.5 inches square and made of plastic. One side has a 5 × 5 peg grid. The other has a circle with a 12-peg circumference. Students stretch rubber bands from peg to peg to form geometric shapes. The Geoboard can be used to study symmetry, congruency, area, and perimeter.

Inchworms™

Plastic Inchworms™ are 1 inch long. Pieces come in six different colors and can be snapped together to make a chain. Inchworms™ are ideal for students who are just starting to learn measurement with standard units, as they provide a transition to using a ruler. They can be used to measure length, width, and height.

Inchworms™ and Centibugs™ Ruler

The Inchworms™ and Centibugs™ Ruler is made of plastic. On one side, each inch of the ruler is marked with an Inchworms™ piece to help students see the units of measurement. The ruler can be used with compatible Inchworms™ to explore using standard units to measure length, width, and height. The other side of the ruler is marked with a Centibugs™ piece to help students explore standard metric units of measurement.

Mini Relational GeoSolids®

Mini Relational GeoSolids® show relationships between surface area, volume, and shape. Solids come in a variety of geometric shapes, including cylinders, cones, triangular prisms, triangular pyramids, cubes, square pyramids, square prisms, and hexagonal prisms.

NumberLine Clock™

The NumberLine Clock™ is a hands-on way to go from counting to telling time. It deepens understanding by visualizing the connection between a number line and time. The color coding helps students recognize groups of 5 minutes. Simply bend the number lines into a circle, connect the ends, put them into the clock face, and snap on the hands, and a clock has been created.

Index

Boldface page numbers indicate a manipulative is used in the **Try It!** activity.

Bar graphs
    create and analyze, 48–49, 82–83
    solving problems on, 50–51, 84–85
Bar Graphs Recording Sheet, 48–**49**, 82
Bookmark Line Plot Recording Sheet, 40–**41**, 74
Centibugs™ Ruler
    estimate length, 22–**23**, 60
    measure, customary and metric units, 24–**25**, 62
    select and use measurement tools, 26–**27**, 64
Coin Tiles
    bar graphs, 48–**49**, 82–83
    solving coin problems, 36–**37**, 73
Color Tiles
    bar graphs, 48
    bar graphs, solving problems on, 50
    compare lengths, 28
    estimate, customary and metric units, 24
    estimate length, 22–**23**, 60
    line plots using inches, 40
    partition rectangles, 12–**13**, 14–**15**, 16, 55, 56
    select and use measurement tools, 26
    whole numbers as lengths on a number line, 30
Cuisenaire® Rods
    bar graphs, creating, 48–**49**
    bar graphs, solving problems on, 50–**51**
    estimate, customary and metric units, 24
    estimate length, 22–**23**, 60
    line plots, solving problems using, 44
    line plots using centimeters, 42–**43**
    measure, customary and metric units, 24–**25**, 62
    select and use measurement tools, 26–**27**
    whole numbers as lengths on a number line, 30–**31**, 69
Customary units
    compare lengths, 28–29, 66
    estimate length, 22–23, 60
    line plots using inches, 40–41, 44–45, 74–75, 78–79
    measure length, 24–25, 62
    select and use tools, 26–27, 64

Data
    bar graphs, 48–49, 50–51, 82–83, 84–85
    line plots, solving problems using, 44–45, 78–79
    line plots using centimeters, 42–43, 76–77
    line plots using inches, 40–41, 74–75
    picture graphs, 46–47, 80–81
Dot Paper, identify shapes, 8–**9**, 52
Rainbow Fraction® Circles, partition circles, 18–**19**, 58
Geoboard
    identify shapes, 8–**9**, 53
    partition circles, 18
    partition rectangles, 12, 14, 16–**17**, 57
    recognizing and drawing shapes, 10–**11**, 54
Geometry
    identify shapes, 8–9, 53
    partition circles, 18–19, 58
    partition rectangles, 12–13, 14–15, 16–17, 55, 56, 57
    recognizing and drawing shapes, 10–11, 54
Hundred Chart, solving coin problems, 36–**37**, 72
Inchworms™
    compare lengths, 28–**29**, 66
    estimate length, 22–**23**, 60
    line plots, solving problems using, 44–**45**, 78–79
    line plots using inches, 40–**41**, 74–75
    measure, customary and metric units, 24–**25**, 62
    select and use measurement tools, 26–**27**, 64
Inchworms™ Ruler
    line plots, solving problems using, 44–**45**, 78–79
    measure, customary and metric units, 24–**25**, 62
    select and use measurement tools, 26–**27**, 64
Insect Recording Sheet, 50–**51**, 84
Length
    compare, 28–29, 66
    estimate, 22–23, 60
    measure, customary and metric units, 24–25, 62
    select and use measurement tools, 26–27, 64
    whole numbers as, on number lines, 30–31, 69

Line plots
    solving problems using, 44–45, 78–79
    using centimeters, 42–43, 76–77
    using inches, 40–41, 74–75
Line Plots Recording Sheet, 44–**45**, 78
Measurement, length
    compare, 28–29, 66
    estimate, customary and metric units, 24–25, 62
    estimate with customary and metric units, 22–23, 60
    recording data of, on line plots, 40–41, 42–43, 44–45, 74–75, 76–77, 78–79
    select and use tools, 26–27, 64
Measurement, time
    telling to 5 minutes, 32–33, 87
    telling to the minute, 34–35, 87
Measurement Recording Sheet 1, 22–**23**, 59
Measurement Recording Sheet 2, 24–**25**, 61
Measurement Recording Sheet 3, 26–**27**, 63
Measurement Recording Sheet 4, 28–**29**, 65
Metric units
    compare lengths, 28–29, 66
    estimate length, 22–23, 60
    line plots using centimeters, 42–43, 76–77
    measure length, 24–25, 62
    select and use tools, 26–27, 64
Mini Relational GeoSolids®
    picture graphs, 46–**47**, 81
    recognizing and drawing shapes, 10–**11**, 54
Money, solving coin problems, 36–37, 73
Nail Line Plot Recording Sheet, 42–**43**, 76
NumberLine Clock™
    telling to 5 minutes, 32–**33**, 70
    telling to the minute, 34–**35**, 71
Number lines, whole numbers as lengths on, 30–31, 69
1-cm Number Lines, 30–**31**, 67
1-Inch Grid Paper, 86
1-Inch Number Lines, 30–**31**, 68
Partition circles, 18–19, 58
Partition rectangles
    into fair shares, 16–17, 57
    into rows and columns, 12–13, 55
    solving problems by, 14–15, 56

Pattern Blocks, 8
Picture graphs, 46–47, 80–81
Picture Graphs Recording Sheets, 46–**47**, 80
Shapes
    circles, partitioning, 18–19, 58
    cone, 10–11, 54
    cubes, 8–9, 10–11, 53, 54
    cylinder, 10–11, 54
    hexagons, 8–9, 53
    identifying, 8–9, 53
    pentagons, 8–9, 53
    prism, 10–11, 54
    pyramid, 10–11, 54
    quadrilaterals, 8–9, 53
    recognizing and drawing, 10–11, 54
    rectangles, partitioning, 12–13, 14–15, 16–17, 55, 56, 57
Telling Time Recording Sheet, 32–**33**, 34–**35**, 87
Time
    telling to 5 minutes, 32–33, 70
    telling to the minute, 34–35, 71

UniLink® Cubes
    bar graphs, 48
    bar graphs, solving problems on, 50
Vocabulary
    bar graph, 48, 50
    bill, 36
    centimeters, 22, 24, 30
    columns, 14
    compare, 28
    cone, 10
    cubes, 10
    customary, 22, 24
    cylinder, 10
    dime, 36
    dollar, 36
    estimate, 22
    fourths, 16, 18
    graph, 44, 46
    halves, 16, 18
    hand, 32, 34
    hexagons, 8
    hour, 32, 34

inch, 22, 24, 28
line, 40
line plot, 42, 44, 46
long, 12
measuring tape, 26
meterstick, 26
metric, 22, 24
minute, 32, 34
nickel, 36
partition, 18
penny, 36
pentagons, 8
plot, 40
prism, 10
pyramid, 10
quadrilaterals, 8
quarters, 36
rows, 14
ruler, 26
thirds, 16, 18
wide, 12
yardstick, 26